INDUSTRIAL ARCH of the SWANSEA R

Fforest Copperworks, Morriston and the Beaufort Bridge, c1790. *Royal Institution of South Wales*

The region which this gazetteer describes is the western half of the south Wales coalfield excluding Pembrokeshire. It also includes the narrow limestone belt which surrounds the coalfield and a few additional sites which are closely linked to this area.

It is, of course, a region with a long history of industrial activity, and of particular importance in the field of nonferrous metallurgy. In recent years the traditional types of heavy industry with which it is so very much associated in the popular mind have contracted sharply and economic activity is now characterised as much by tourism and services as by metal processing and coal mining.

It might be thought, in an old-established industrial region, where many of the original industries have gone into decline, that there would be an abundance of sites of major IA significance. But in fact, in recent years widespread reclamation projects have been initiated with great, if not excessive, enthusiasm, often born as much of political motives as of environmental considerations. There are thus fewer outstanding sites to be seen today than one would like or would have expected, or indeed than there were some twenty or thirty years ago.

Nevertheless it has been no problem to assemble the sites listed in this book. All of them are of historical interest and many are visually impressive. Frequently, too, they are in fine surroundings; for while south Wales has been heavily industrialised and is still densely populated, this has normally been confined to the lower lying land along the coast or on the valley floors, with the result that the hilly country remains largely uncontaminated by urban sprawl. It is a region that has much to offer the visitor, and it is hoped that this little book will help to reveal some of the features in its industrial history.

Stephen Hughes
Paul Reynolds

HOW TO USE THIS GAZETTEER

The gazetteer is divided into sections which correspond to different industrial categories. In each section there is first a brief introduction to the history of that particular industry as found in the Swansea region; there then follow brief descriptions of some of the more interesting sites connected with the industry.

Readers who wish to use the gazetteer as an aid to exploring a particular part of the region should turn to the map in the centre of the book. This shows all the sites which are described in the book. The numbers against each one of them correspond to the site numbers in the text. The map therefore serves as a geographical index to the text.

The heading to each entry contains certain coded information in addition to the name of the site and its 8-figure grid reference.

ACCESS CODES

★★   A major site, fully visible inside and out by the public from places to which they normally have access. In the case of a site where there is nothing significant internally, this code can also indicate a site in private hands which can be viewed adequately from a public place.

☆☆   A major site which can only be viewed satisfactorily with the permission of the owner.

★   A site of less importance which can be inspected freely by the public.

☆   A site of less importance which can only be inspected satisfactorily with the permission of the owner.

All sites with ☆☆ or ★★ codes are important sites which especially merit official protection.

INFORMATION CODES

These indicate that further information is held by the Royal Commission on Ancient Monuments (Wales) or by the National Monuments Record.

N   Notes.
I   Illustrations (prints, photographs, etc.)
PS   Partial survey.
S   Full survey.

The address for both the Royal Commission and the NMR is Edleston House, Queen's Road, Aberystwyth SY23 2HP.

CONSERVATION STATUS

SC   Scheduled.
LS   Listed.
NT   In guardianship of The National Trust.
CA   Conservation Area.
☐   No statutory protection.

Smith's Canal Tunnel, Swansea [84].

COAL MINING

Among the industries of south Wales the one that is basic to all the others is the coal industry. It was the first industry in the region and even today the image that outsiders have of south Wales is based on the traditional mining community.

The earliest reference to coal mining in the Swansea area occurs in 1306 and by the 16th century Swansea had a flourishing export trade in coal. By the 17th century it was probably the third largest coal port in the country after Newcastle and Sunderland.

The existence of cheap and readily accessible coal was a major factor that contributed to the location of the metal smelting industries in the Swansea area, and in turn these industries stimulated demand for coal. Output in the lower Swansea valley probably peaked at the end of the 18th century. During this period the industry was dominated by two industrial dynasties, the Morris family on the western side of the valley and the Townsend/Smith family on the east. Initially both families ran integrated smelting and mining businesses but by the end of the century they had pulled out of smelting and preferred to concentrate their efforts on supplying coal to independent smelting companies.

The geology of the coalfield is such in the Swansea area that in many cases coal was accessible by level or drift. One of the earliest methods of working coal was by bell pits and examples can be found in the Clyne valley and in the Cwmllwyd woods to the west and north-west of Swansea. But by the beginning of the 18th century – and probably before – it had become necessary to sink shafts for the coal. These pits required to be pumped and as early as 1717 a Newcomen engine is known to have been at work on an unidentified site on the eastern side of the Tawe.

Another means of extracting the coal, which appears to have been pioneered at Swansea, was by boat level: the earliest example known was built by Robert Morris at his Clyndu Level, the site of which is now covered by a large traffic island at the south end of Morriston. Two other local underground canals may also have pre-dated the famous Worsley navigation levels of 1759. The best surviving example at Swansea, however, is comparatively late, at the Rhydydefaid colliery in the Clyne valley which is dated to about 1840.

One of the specialities of south Wales is anthracite and within the coalfield this is found only in Pembrokeshire and around the north-western edge of the coalfield in a belt stretching from the Gwendraeth valleys to the head of the Neath valley. Until the very end of the 19th century demand for this fuel was low, but then the export market opened up, especially to the continent where anthracite was in demand for domestic stoves. This meant that the anthracite coalfield developed later than the steam coalfield and did not peak until 1934, compared to 1913 for the coalfield in general. Swansea was well placed to benefit from this growing trade.

Substantial remains of the coal industry tend to be rather few considering the importance it once had. Many sites have been cleared and landscaped. However, throughout the region less spectacular evidence of coal mining can be seen in the form of collapsed level mouths, partially grassed-over spoil heaps or small groups of tumbledown buildings.

1 RHYDYDEFAID BOAT LEVEL
★   N I   [SS 6038 9218]   ☐

A flooded cutting in the Clyne Valley Country Park probably represents the surface remains of a short-lived underground colliery canal. It was located at the end of a branch of the Oystermouth Tramroad and built c1840 by Sir John Morris II. Nearby can be seen the remains of later brick-built engine-houses and an early 20th century steam winding engine, left to moulder among the trees.

2 PWLL MAWR (GREAT PIT)
★   N I PS   [SS 6772 9629]   ☐

Sunk c1772 by John Smith on the site of a borehole made by Chauncey Townsend c1770. It worked the Great or Six Foot Seam. The shaft was 450' (137m) deep and worked until 1828 when a disastrous explosion led to its closure. In 1872 it was acquired by Evan Matthew Richards and re-opened in 1881. The present pumping house (of which one monumental rusticated wall remains) dates from this period. It was finally closed c1893. The course of Townsend's wooden railway can be seen on the hillside beside the pit.

3 GWERNLLWYNCHWITH ENGINE HOUSE
★★　　N I S　　[SS 6975 9796]　　SC

This ivy-grown crumbling ruin is a very important monument, for it is probably the earliest engine house still existing that housed a rotary motion (as opposed to reciprocating) steam engine. It was built by the coalowner John Smith between 1772 and 1782 and was out of use by 1786. It is commonly supposed that no rotary steam engines were used before Watt's improved engine, which was first applied to coal winding in 1784.

4 SCOTT'S PIT, LLANSAMLET
★★　　N I S　　[SS 6972 9830]　　SC

Sunk in 1817-19 by a London solicitor, John Scott. It proved unremunerative and Scott and his partners sold it to the local coalowner, C H Smith, who worked it until c1842. In 1872 the engine house was recommissioned but only for pumping and draining the newly developed Cae Pridd colliery. It remained in use for this purpose, intermittently and with a succession of owners, until 1930. The main surviving feature is the engine house which was restored in 1976-80 and taken into the possession of Swansea City Council. In addition there are the foundations of a haystack boiler (c1820) and of a Cornish boiler (c1872), the foundations of a stack for furnace ventilation and the site of the 500' (150m) deep shaft, now capped in concrete. A tramroad ran from Scott's Pit to a shipping place at White Rock [105].

5　　　　　　　　　　　　　　　　GARTH PIT
★　　N I　　[SN 6911 0026]　□

Sunk 144' (44m) to the Church or Four Foot Vein in 1834. The lower part of an engine house of this date, built of Pennant sandstone, still remains. Tramroad sleeper blocks can be followed along the track leading to the River Tawe where one bridge abutment remains of the bridge which carried the tramroad over to the Swansea Canal.

6 FELINDRE COLLIERY ENGINE HOUSE
★★　　N I S　　[SN 6542 0394]　□

A steam-pump engine house of rusticated Pennant sandstone built in 1879 by the colliery proprietors Cory & Yeo. As part of the same project it was also intended to build a railway several miles long to serve this pit and the Daren Colliery [SN 6550 0515] in order to connect them to the tramway in Cwm Clydach, but the whole enterprise was abandoned, possibly because the sinkers encountered quicksand. The engine house is now a dwelling house overlooking the Lower Lliw reservoir.

7　　　　　　　　　　　　　CRIMEA COLLIERY
★★　　N I　　[SN 7570 0730]　□

The beam-engine and winding houses remain substantially intact together with the railway formation which lead to the remains of a tipping-stage beside the Swansea Canal. It was sunk in 1854 – of course – and lasted until 1862.

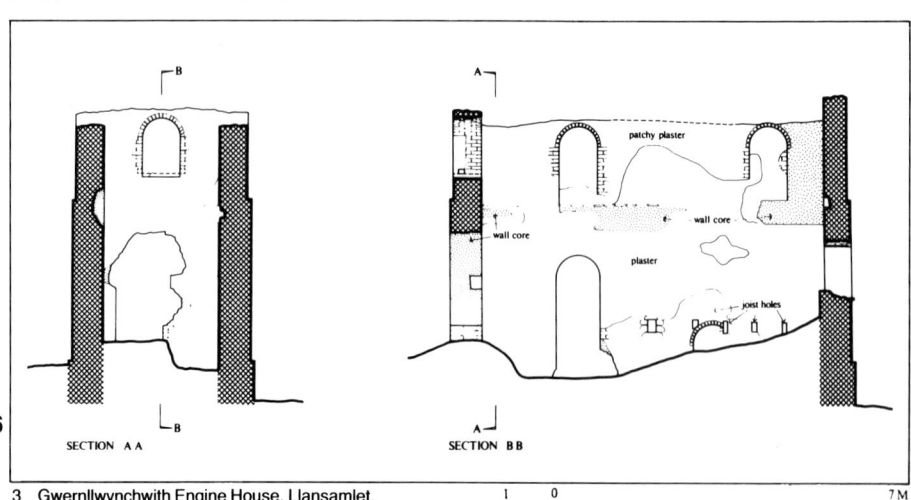

3　Gwernllwynchwith Engine House, Llansamlet.

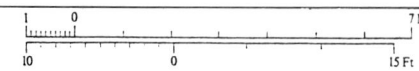

8 YNYSGEDWYN COLLIERY VENTILATION SHAFT
★   [SN 8106 1113] ☐

A masonry and brick tower built to house a Guibal fan for extracting air from Ynysgedwyn Colliery.

9 HENLLYS VALE COLLIERY
★   N I PS   [SN 7617 1360] ☐

An anthracite drift on the very edge of the coalfield which operated c1890-1930. The main surviving feature is a tall brick-built chimney stack. Close by is a bank of limekilns [19] which must have burned coal from the drift. The site is best approached by foot from Brynhenllys bridge [SN 7557 1257] along a path which follows the former railway line (one mile, rough in places). Between this path and the River Twrch are sections of a leat which fed water to Lower Brynhenllys colliery [SN 7550 1215] for operating pumps and screens. This colliery closed in 1955 and some confused modern remains still exist.

10 CEFN COED WINDING ENGINE
★★   N   [SN 7853 0333] ☐

When it was sunk in 1926 Cefn Coed was the deepest anthracite mine in the world with two shafts of over 800 yds (732m) depth. Two engines were installed to wind the cages up and down the shafts. The colliery itself closed in 1968 but the surface buildings and winders remained in use for the adjacent Blaenant drift until 1978. They were then acquired by West Glamorgan County Council and developed as a museum.

Among the items on display are:-
—the single surviving winding engine (Worsley Mesnes, Wigan, 1927).
—a suite of Lancashire boilers.
—the downcast (No 2) shaft headgear.
—the secondary winder.(Cubitt, Rhondda).
—the compressor house containing one of the two original compressors (Bellis & Morcom, Birmingham).
—the stack and flue of the boiler house.
A former NCB locomotive has been installed in the car park and now serves as a climbing frame.

11 GENWEN ENGINE HOUSE
★★   N I S   [SS 5456 9958] ☐

Built by George Warde in 1806 to replace an earlier engine house of 1766 erected by Chauncey Townsend. The colliery fell out of use from the mid-1820s until 1837 when the Llanelly Copperworks Co installed a new engine and deepened the pit. They continued to work it until c1867. The pit was reopened and deepened yet again in 1898 and worked until 1907 when an explosion caused the seam to ignite. Pumping ceased in 1908, the water rose and the colliery was flooded out. The engine house is substantially as erected by Warde but with later alterations. It is heavily festooned with ivy and in a derelict condition.

12 PENPRYS PIT
★★   N I S   [SN 5400 0181] ☐

Developed c1840 by the Llangennech Coal Co and worked until c1885. Intermittent working followed until final closure in 1908. The engine house stands to its full height, although roofless.

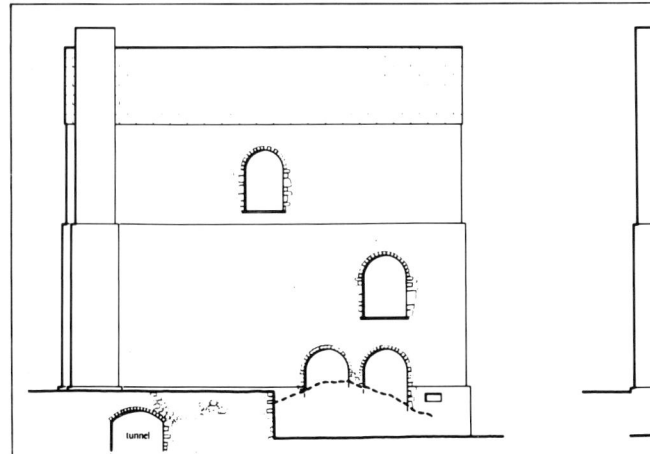

RECONSTRUCTED NORTH ELEVATION

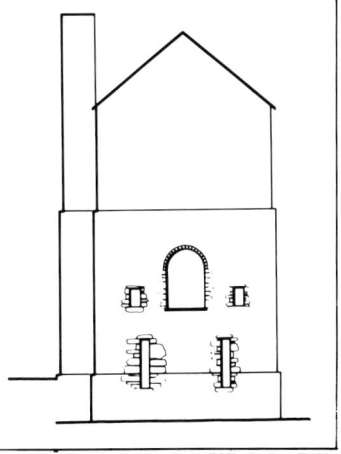

RECONSTRUCTED WEST ELEVATION

13 ST DAVID'S PIT
★   **N I**   [SN 5395 0135]   ☐

Sunk by the Llangennech Coal Co in 1827 and linked by the Llanelly Railway [146–149] to the New Dock at Llanelli [73]. Coal was struck at 660' (200m) in 1832, making the pit the deepest at the time. Monumental but derelict fragments of masonry remain. The 1833 railway incline can be followed as a metalled footpath.

14 PWLL COLLIERY ENGINE HOUSE
★   [SN 4756 0096]   ☐

Pwll (or Pool) colliery dates back to 1765 and was the site of an early engine (c1769). The present truncated engine house, however, belongs to later operations, either by M J Roberts (1825) or—perhaps more likely—by Mason & Elkington, the copper smelters of Pembrey [33] who took the colliery in 1865. Abandoned c1880.

For further sites associated with coal mining, see entries 59, 83, 109 and 120.

LIMESTONE & OTHER QUARRYING ACTIVITY

Limestone forms a thin belt all round the coal measures of south Wales. It is particularly prevalent in the Gower peninsula where this belt becomes broader than in other parts. Limestone has been valued and worked for many years for a number of different purposes. In its raw state it is used as a flux in iron smelting, and after calcining and being reduced to lime it forms an agricultural fertiliser and was used for making mortar and for limewashing the walls of houses. Throughout the limestone belt one can find quarries from which limestone was extracted for these purposes, very often with an associated bank of limekilns. The proximity of the coalfield ensured easy supplies of coal, and the limekilns were able to utilise culm, the small of anthracite, for which otherwise the demand was weak.

Because of the combustible nature of slaked lime, the normal practice was to burn lime as close as possible to the point of use. This resulted in the building of kilns right away from the source of limestone, e.g., in Swansea itself. An important market for limestone was in Devon and Cornwall where the stone does not occur naturally. Until chemical fertilisers became widespread limestone was shipped across the Bristol Channel, especially from Gower, and was calcined on the spot using coal which was also shipped over from south Wales. This trade, and the quarries which supplied it, lasted until c1900.

Another form of limestone which was in demand in the 18th and 19th centuries was rottenstone. This is a very friable decayed limestone which was quarried at Cribarth at the head of the Swansea Valley. Rottenstone was used in the local copper and tinplate industries for polishing metal sheets and also as a domestic abrasive.

Between the coal measures and the limestone there runs a belt of millstone grit along the northern rim of the coalfield. Within this belt of gritstone, quartzite or silica is commonly found. It has been exploited for the manufacture of firebricks, in particular at Pontneddfechan and at Mynydd-y-Garreg, near Kidwelly. At Penwyllt silica sand was quarried for this purpose and made into firebricks.

The construction industry was also an important user of stone. The coal measures are overlaid by Pennant sandstone, a dull, if fairly easily worked building stone. Quarries can of course be found throughout the region and none of them are of any outstanding interest. Two of the larger examples are Rosehill [SS 643 935] at Swansea, from which much of Victorian Swansea was hewed, and the quarries around the southern slopes of Kilvey Hill [SS 67 93] from which was obtained stone for Swansea docks.

15 PWLL-DU QUARRY
★   [SS 5700 8630]   ☐

Limestone was being quarried here on the coast of Gower by 1650, and then it was said to have been worked since 'tyme out of mynde'. The existence of a large block of easily worked limestone close to a beach meant that an intensive trade was carried on with Devon and Cornwall until quarrying ceased in about 1900. Note the slides down which blocks of stone were lowered to the beach.

GOWER LIMEKILNS

Many small limekilns intended to supply a purely local need for lime for agricultural purposes can be found throughout the Gower peninsula. Most of them probably fell into disuse early in the present century. Two typical examples are:

16   VENNAWAY
★   [SS 5645 8958]   ☐

Easily spotted on a main tourist approach to Gower.

17   LLANGENYDD
★   [SS 4325 9075]   ☐

Double-hearthed kiln standing about 12′ (3·6m) high, stated to have been worked to c1930. A quarry from which the limestone was obtained can be seen nearby. The provision of two hearths was to allow for different wind directions and can be found in other Gower limekilns, e.g. Great Tor ★ [SS 5305 8790]. There are many limekilns in Gower, and a perusal of the 6″ Ordnance Survey map will reveal the location of others.

18   HAFOD LIMEKILN
★   N I S   [SS 6610 9499]   **LS**

The last surviving example of the 54 limekilns that once stood alongside the Swansea Canal, although fragmentary remains survive at Pontardawe, Cae'r-lan and Yard Bridge. Limestone was brought in by the Oystermouth Railway and Swansea Canal from Mumbles and anthracite came in by canal from the upper Swansea valley. This mid-19th century kiln differs from others on the canal in being built above canal level and not into the actual formation, and also because it was used to supply lime for building rather than agricultural use. Copper slag was dumped to provide a ramp to the kiln top.

19   HENLLYS VALE LIMEKILNS
★   N I S   [SN 7620 1362]   ☐

A bank of four massive limekilns built of stone and brick and probably late 19th century. There appear to be two different phases of construction. Coal was no doubt obtained from the nearby Henllys Vale colliery [9], and limestone was brought in from quarries on Cefn Carn-Fadog by a tramway which can still be seen. The best approach is by the footpath which follows the track of the railway by which the product was taken out [see 9].

20   CRIBARTH MOUNTAIN LIMESTONE SILICA SAND & ROTTENSTONE QUARRIES
★★   N I S   [SN 8325 1435]   ☐

More than 30 limestone quarries on Cribarth were made accessible to the Swansea Canal by means of an eventual total of 18 inclined planes and some 10 miles of railway built in all periods between 1794 and the 1890s. The main period of activity coincided with the boom in the local anthracite iron industry (1837–c1860) and these quarries and railways are the main memorial to the frenetic activity of that period. Silica sand for furnace linings was also worked in quarries around the mountain top, as was rottenstone, the impure limestone shale used in polishing copper and tinplate. An unfinished tramroad built in the 1820s for one mile to the north of the limestone outcrop is one of the best sources for understanding how such early horse-worked railways were built. There are remains of limekilns on and around the mountain top.

21   CARNAU GWYNION LIMEKILNS
★★   N I S   [SN 9147 1450]   ☐

In and around a walled enclosure some one square kilometre in extent stand the remains of no less than 171 limekilns. There are several such concentrations of kilns on the southern edge of the Great Forest of Brecon to which the commoners came from local farms to burn lime on the limestone outcrop itself. Many of the kilns date from before the enclosure of the Forest in 1819 and nearly all of them were very crudely constructed of unmortared limestone blocks. The remains are inevitably ruinous.

22   TY BONT LIMEKILN
★   N I S   [SN 8107 0188]   **LS**

Accessible from the A465 road at Ynysarwed by an original stone-arched bridge over the Neath Canal. As with nearly all Welsh canalside kilns, this was built into the side of the canal on the valley side so that its top was at charging level.

23   DINAS SILICA MINE
☆☆   N I   [SN 9169 0797]   ☐

Quartzite was mined underground here, following one bed over an area of 1000 yds by 500 yds (915m x 457m). The roof was supported by pillars of quartzite left standing. The silica rock was discovered in the 1780s and work was being pursued enthusiastically by 1807 when a tramroad, Dr Bevan's, was built to take the product to the Neath Canal. It was in 1823 that the renowned Dinas firebrick was invented by W W Young, its constituents being silica with some added lime. This refractory brick for lining furnaces became world-famous, causing the word 'dinas' to pass into the German and Russian languages as a term for 'firebrick'. The silica mine continued to be worked until 1964, latterly by RTB. Its closure was associated with the decline of the open-hearth steel process. The silica was made into bricks at Morriston.

24 LLANDYBIE LIMEKILNS
☆☆ I [SN 6150 1670] **LS**

There is a long history of quarrying and lime-burning in this part of Carmarthenshire (as was), but it was the opening of the Llanelly Railway [146–149] in 1857 that led to a major increase in activity. The surveyor and architect, R K Penson (c1815-1885) was responsible for the quarries being exploited on an industrial scale and he designed the imposing bank of six kilns at Cilyrychen which remain, although no longer in use. Penson's original bank of 1856 was subsequently extended and by 1900 there were nine kilns in production. Penson was primarily an ecclesiastical architect who did a good deal of work in the St Davids diocese: his gothic limekilns are quite in keeping with this style of architecture.

25 FOEL FAWR LIMEKILNS
★ N I PS [SN 7270 1920] ☐

A road across the Black Mountain was built in 1819 to supply coal to the upper Towy valley. It also made it possible to start working the limestone on a large scale. Kilns can be seen close to the A4069 road as it starts to descend into Llangadog.

For further sites incorporating limekilns, see entries 88, 89, 110, 115 and 120.

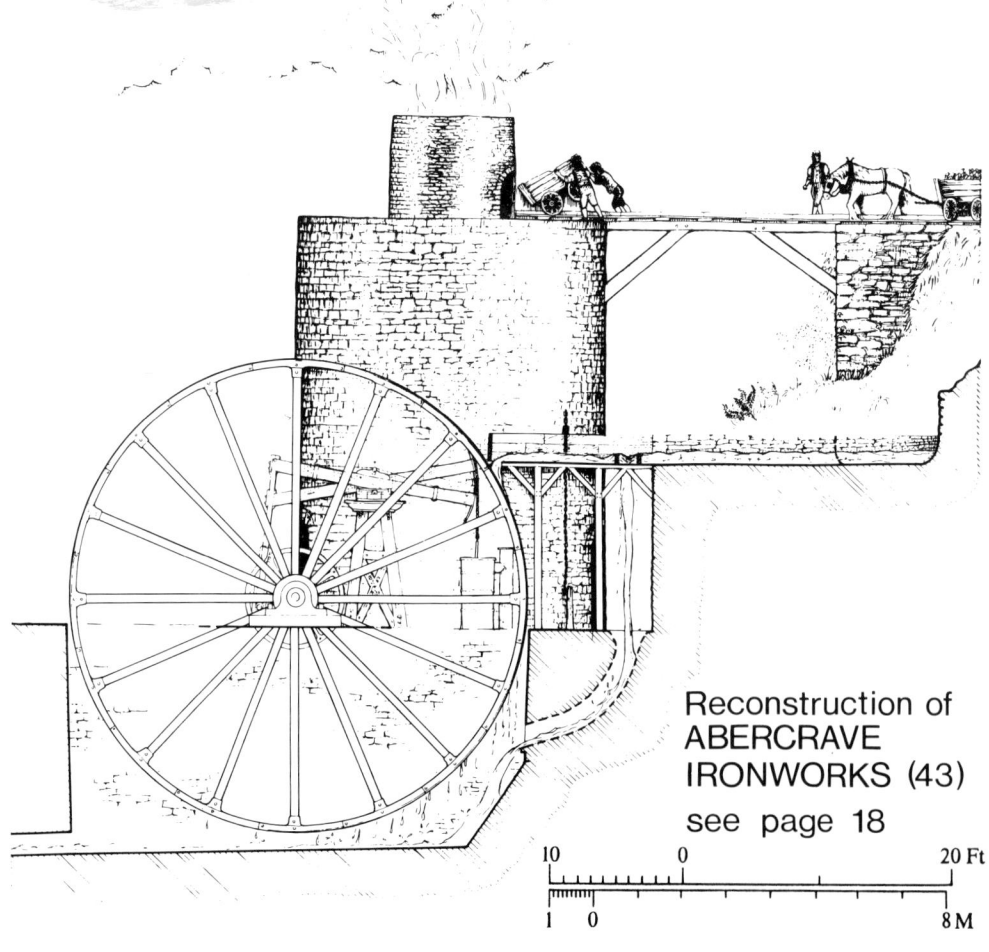

Reconstruction of
ABERCRAVE
IRONWORKS (43)
see page 18

COPPER SMELTING

It was the presence of cheap and abundant coal that led to the establishment of another of the major industries of south Wales, nonferrous metal smelting. In its day about **90 per cent** of Britain's copper smelting capacity was located within 20 miles' radius of Swansea and there were also smelteries processing zinc, lead, silver and other metals.

The industry first appeared in the region in 1584 when the German, Ulrich Frosse, established himself at Aberdulais, near Neath, possibly on the site now owned by the National Trust. Neath remained the main centre of the industry during the 17th century but in the 18th century the centre of the industry moved to Swansea, largely because of the latter's superior harbour. Swansea was an excellent location for the copper industry. It had plenty of easily worked, and therefore cheap, coal, and just across the Bristol Channel was Cornwall, the source of copper ore. Because copper smelting required a considerably larger proportion of coal compared to ore, it made more sense to take the ore to the coal rather than *vice-versa*. Vessels bringing the copper ore to Swansea could also take on a return cargo of coal which was in demand in Cornwall, not least to power the engines that raised the water from the copper mines.

The first copperworks in the lower Swansea valley was the Llangyvelach works, set up in 1717 by John Lane of Bristol. As manager Lane brought in a young man from Shropshire, Robert Morris, who was subsequently to establish one of the major industrial dynasties of Swansea with interests in copper smelting and coal mining. Lane's works was the first in a succession of copperworks and over the next century a total of thirteen were built on both sides of the valley between Swansea and Morriston.

But while copper smelting brought prosperity to Swansea, it also created severe environmental problems. Dense clouds of smoke, laden with sulphuric acid billowed across the valley and blighted the sides of Kilvey Hill, and great quantities of copper slag were produced as an unwanted by-product and deposited all over the valley to form vast unsightly tips. It is only in recent years that the last of them have been removed.

As the 19th century advanced Swansea started to look further afield for supplies of ore. Cornwall was supplemented first by Anglesey and then, as the native ores started to be exhausted, by Cuba, Chile, America and Australia. This foreign trade is perpetuated in the names of public houses in Swansea such as the Cape Horner, the Cuba or the Mexico Fountain.

But even though the copper industry appeared to be well established at Swansea the seeds of decay had already been sown and from about 1870 the industry was in decline. The needs of mass electrification led to a surge in demand for copper which Swansea could not hope to meet, dependent as it was on imported ores. The trend was increasingly to smelt at the point of production and ship the refined copper to the point of manufacture. Swansea smelted its last copper in 1921, although the processing of copper was to continue until 1980, latterly as a very minor component of the local economy.

Many of the copperworks have been entirely demolished and at the others the remains are only fragmentary. The tips have all been cleared and used in airport or motorway construction. Perhaps the most abiding inheritance of Swansea's copper smelting past are the moulded blocks of shiny green-black slag which are to be found in walls all over the city, a favourite use being for coping stones.

26   LLANGYVELACH COPPERWORKS
★   N I   [SS 6613 9595]   ☐

The first of the Swansea copperworks, established in 1717 by John Lane of Bristol. No remains survive above ground and the site is now occupied by the junction of Cwm Level Road and Neath Road to the north of Landore viaduct [127].

27   HAFOD COPPERWORKS
☆☆   N I PS   [SS 6627 9511]   **LS** 11

Established by John Vivian in 1810. The site was adjacent to the Swansea Canal and the low brick arches in the wall of the works mark the former dock entrances. It remained in the hands of the Vivian family until 1924. By then only two copper smelting firms remained in the Swansea valley, Vivians and Williams Foster [28], and they were refining rather than smelting ore. Both firms were acquired by British Copper Manufacturers who in turn were acquired by ICI in 1928. Copper rolling continued (through the ICI subsidiary, Yorkshire Imperial Metals) until closure in 1980. The site was acquired by Swansea City Council and rapidly cleared. Certain features were retained, possibly to form the nucleus of an industrial museum, and these include:-

—engine house of 1860–62 (with dated plaque) which formerly housed an engine to drive the copper rolling mills ☆☆ **N I LS.**

—engine house of 1910, built to house a surviving Musgrove uniflow engine which drove the copper-rolling mills in succession to the earlier engine ☆☆ **N I LS.**

—locomotive shed on a terrace by the river, built for the first standard-gauge Garrett locomotive in Britain.

—copper slag abutment and masonry pier which supported the tramway carrying waste away from the works ☆☆ **N I S LS.**

—late 19th century offices, now rendered in white and used as a social club.

For Hafod limekiln, see entry 18; for Vivianstown, which housed workers at the Hafod, see entry 154.

28　　MORFA COPPERWORKS
☆☆　　**I PS**　　[SS 6607 9517]　　**LS**

The Morfa copper-rolling mills were built by the Cornish firm of Williams, Foster & Co in 1828. The surviving large shed is on this site and may incorporate (or be?) the original building. Originally copper was brought in here from the Rose works, but smelting probably started at Morfa in 1835. The site is adjacent to the Hafod works [27] of Vivian & Sons with whom they amalgamated in 1924. The two works were then combined and worked as one until closure in 1980. Surviving features include:-

—the main works building with pantile roof, parts at least of which probably date from 1828 ☆☆ ☐.

—mid 19th century laboratory building ☆ **LS**.

—canteen (formerly the power-house) with an aisled interior and a clock turret on the roof ☆☆ **LS**.

—red-brick office block, c1900.

Adjacent to the remains of the smelting works are two other important features:-

MORFA QUAY ★　[SS 6644 9535]　　**LS,SC**
Built 1835 for unloading copper ore into the works from river-borne vessels. It consists of a stout timber frame filled with brick and carried a broad-gauge railway.

MORFA BRIDGE ★★　**I PS**
[SS 6641 9543]　　**LS,SC**
Iron and timber bascule bridge built 1909 to link the Morfa and Upper Bank works so as to carry the waste from Morfa to tips on the eastern side of the river. Formerly raised by hydraulic power but now fixed.

29　　　　　　　　　　WHITE ROCK
★★　　**N I PS**　　[SS 6627 9476]　　☐

The third oldest of the Swansea copperworks, set up in 1736 by a partnership from Bristol at a time when copper smelting was switching from blast-furnace to reverbatory furnace technology. The site of the 'Great Workhouse' of this period is known. The location was attractive for a number of reasons: the Mansels' Great Coal Road terminated here in a 17th century river dock and wharf which has recently been re-excavated. The feed to an earlier water mill also existed, and this was subsequently greatly extended to form a leat which reached to the far side of Kilvey Hill. The road from Llansamlet was later transformed into a wagonway by Chauncey Townsend and then in 1783-85 Smith's Canal [84] was driven through the densely developed site in a cut-and-cover tunnel with openings to provide direct access for coal supplies. In 1870-71 lead and silver smelting were introduced, and this resulted in the building of a brick condensing flue up the side of Kilvey Hill and an inclined railway to remove the spoil. Part of the latter was supported by a stone arch that also housed two flues and a chimney, and this still survives. A range of 19th century stone-built wharves remains along the river frontage with decks made of cast blocks of copper slag. The early White Rock quay has collapsed and been removed, but at its southern end are the remains of a chute leading down to wharf level. Above this point was the terminus of Scott's tramroad [105].

White Rock closed in 1924. The site was almost completely cleared in the 1960s. What survived was designated as the White Rock Industrial Archaeology Park in the 1980s.

30　　　UPPER BANK COPPERWORKS
★　　**N I**　　[SS 6656 9530]　　☐

Founded c1757 by Chauncey Townsend and supplied with coal by his wagonway. Initially lead and zinc were smelted but it was converted to copper smelting in 1777. During the period 1838-42 it was occupied by the Muntz

29 White Rock Copperworks, before 'reclamation'.

Patent Metal Co and it was here that Muntz Metal, or Yellow Metal, was invented, a zinc-copper alloy used for sheathing the bottoms of ships. Subsequently it belonged to the Grenfell family until taken over by Williams, Foster in 1893. It closed in 1928.

The re-roofed buildings of Upper Bank remain and are occupied by a firm of plastic brush manufacturers. Note the gable-ends pierced with large circular openings and the quay walls on the river frontage.

Between Upper Bank and White Rock is the desolate site formerly occupied by Middle Bank [SS 6633 9493], also built by Townsend in 1755 to smelt lead and used for copper smelting, 1765-1924.

For housing occupied by workers at Upper Bank and Middle Bank, see entry 155.

31 PENCLAWDD COPPERWORKS
★   N   [SS 5470 9585]   ☐

Copper was smelted at Penclawdd from some date prior to 1788 until 1811 and again between c1848 and 1868 with interruptions. Subsequently the site was used for lead and silver smelting and for tinplate manufacture. John Vivian operated here (1800-1811) before establishing his Hafod works in Swansea [27]. One wall of the first smelting works still stands together with an archway that dates from the 1848 rebuilding by Low's Patent Copper Co, as a plaque under the ivy records. A tramroad embankment to the east of the works, which brought in coal from about half a mile away, may belong to the same period. Blocks of copper slag feature prominently in its construction. The remains of the works own sea-dock are alongside. The ruins on the works site belong mainly to the modern cockle processing plant.

32 CROWN COPPERWORKS
☆   [SS 7302 9663]   ☐

Erected at some date prior to 1797 by the Rose Copper Co of Birmingham. It later passed to Williams, Foster of the Morfa works in Swansea [28] who owned it until 1866 when it was sold to a new owner. Soon after this copper-smelting ceased and the works was later adapted for zinc smelting. In the 1920s patent fuel was manufactured on the site. The remaining buildings, constructed largely of blocks of copper slag, are now engulfed in a flood of wrecked cars which makes interpretation of the site difficult.

The original water-supply pond survives on the western side of the Neath by-pass [SS 7295 9679] and water was conveyed over the Tennant Canal by an aqueduct.

33 PEMBREY COPPERWORKS
☆   N   [SN 4490 0035]   ☐

The last of the copperworks in the region to be established, it occupies a site beside the dock at Burry Port [75]. It was erected by the Birmingham firm of Mason & Elkington in 1849, and a lead and silverworks were added to the north in 1853. Mason & Elkington sold out to Elliott's Metal Co in 1884, and the works was closed in 1912. The building is now used by a chemical company, externally almost unaltered. Construction is of copper slag, stone and brick.

For sites on which copper smelting preceded a later industry, see entries 51 and 54.

48 Banwen Ironworks.

OTHER NONFERROUS SMELTING

The successful establishment of copper smelting in the Swansea region led to the location of other nonferrous smelting industries in the area, in particular lead and zinc (some of the copperworks also produced brass).

Lead was being smelted at Neath in the 1690s. Smelteries were erected near Llanelli in about 1754, and at Swansea in 1755 when Chauncey Townsend built his Middle Bank works, followed in 1757 by the Upper Bank works. None of these ventures proved to be very lasting and it was not until well into the 19th century that lead smelting achieved any great importance, mainly at White Rock where lead replaced copper after the whole works was taken over by Williams, Foster (c1871-1924). An important by-product of lead smelting was silver and this too was produced at White Rock and also at the short-lived Landore silverworks (1853-55).

Zinc or spelter was the most important smelting industry in Swansea after copper, and as copper-smelting went into decline from about 1870, so zinc took on an increasing importance. The first local attempt at zinc smelting was by Townsend at Upper Bank (1757) but without apparently much success. The invention of the copper/zinc alloy, Muntz Metal in 1832, followed by that of the galvanising process led to a major increase in demand. Starting in 1841 a number of zinc smelting works were opened in the Swansea valley. The first of these, the Morriston spelter works of Vivian & Sons, the copper company, was on the site of the abandoned Birmingham copper works (1841-1923). It was followed by a string of four alongside the South Wales Railway at Llansamlet and a sixth in the docks area. Zinc smelting survived at Swansea until 1971 when the last smelting works of any description in the lower Swansea valley, the Imperial Smelting Corporation's Swansea Vale Works, closed.

The process normally used for zinc smelting in the Swansea area was the Belgian or Silesian process. It was as a result of the adoption of these processes that colonies of Belgian and German workmen were attracted to Swansea. Less common was the English method of smelting by downward distillation which required much larger quantities of coal. Until recently a rare—if not unique—example of a building designed for this process survived at Loughor.

34 CLYNE WOOD ARSENIC AND COPPER WORKS
★★ N I   [SS 6147 9085] ☐

The only nonferrous smelting works of which substantially complete remains survive from the fifty or so works that made south-west Wales the world centre of this industry. The site is in dense woodland in the Clyne Valley Country Park, and its survival is largely due to its early failure and abandonment. It was built between 1825 and c1840—possibly in 1837—for a Cornish company and closed in 1841. It was then re-opened in 1844 by Henry Kingscote and by 1852 was being run by the Jennings family who in 1860 transferred their activities to a site with better transport links and coal supplies.

The remains of several pipes, flues and furnaces survive. A large uphill flue terminates in the picturesque 'Ivy Tower' [SS 6133 9081], which in reality is the stump of the main chimney-stack of the works to which have been added battlements, an internal staircase, a gothic window and a door! There may originally have been a condensing chamber in the base of the stack.

The Clyne Valley Canal of 1799-1803 skirts the lower part of the site and what appears to have been a dock or inlet may have allowed small tub-boats from a coal level upstream to enter the works.

35 MOND NICKEL WORKS, CLYDACH
★   [SN 6960 0135] ☐

Established 1902 by the Mond Nickel Co to refine nickel matte from Canada by means of the nickel carbonyl process. This was devised by Sir Ludwig Mond (1839-1909) and involved the use of carbon monoxide. Ownership passed to INCO in 1928 who still operate the plant. 'The Mond' is the last nonferrous works in the Swansea valley and occupies an extensive site between the canal and the River Tawe. A walk along the towpath permits limited inspection of the plant currently on the site. The main frontage seen by the public is a modest two-storey block built of brick with round-headed openings. Buildings in a similar style—in many cases no longer in use—remain within the works, although in recent years much new plant has been installed. Opposite the main entrance, in a paved enclosure, the statue of Sir Ludwig contemplates the village.

1754-55 on a site close to the Loughor estuary, the first smelting house on the Llanelli coalfield. It does not seem to have been very successful and probably closed in about 1770. Much of the original structure was removed during the construction of the Llanelly Railway's extension to Pantyffynnon [146, 147] in 1837-39, but some stone walls still stand to roof height.

36   PENCOED LEADWORKS
☆   **N I S**   [SS 5606 9954]   ☐

A 'smelting house for lead' was erected in

For further sites at which nonferrous metals (other than copper) were smelted, see entries 29, 32 and 33.

IRON AND STEEL

In the classic period of Welsh industrial history, say between 1750 and 1850, there were two great metallurgical districts in south Wales, the copper zone which was centred on Swansea, and the iron zone which stretched all the way along the blaenau, or heads of the valleys, from Hirwaun to Brynmawr. By the middle of the 19th century, however, the inland iron industry was in decline: the local ores were nearing exhaustion and the new steel-manufacturing processes which were being introduced required non-phosphoric imported ores. The ferrous industry moved from the blaenau down to the coastal strip so as to be near the harbours and so was born the steel industry of Llanelli, Swansea, Briton Ferry, Port Talbot, Cardiff and Newport.

But well before the rise of Merthyr there were little rural forges and furnaces. Examples include Melincwrt, Llandyfan and Clydach. Originally these worked with charcoal and went over to coke in the late 18th century. One such furnace was at Ynysgedwyn at the top of the Swansea valley which had developed into a major undertaking by the beginning of the 19th century. At this period it was importing coal from lower down the valley because the local anthracite could not be used for iron-smelting, but in 1837 a major breakthrough occurred when it was shown to be possible to use anthracite for furnace work. This led to a rash of new ironworks in the anthracite belt, all eager to exploit the new process, and for a short time anthracite iron looked set to present a real alternative to steel. Of the most important of the anthracite iron works, Ystalyfera and Ynysgedwyn, almost nothing remains. There are, however, good remains of smaller works at Banwen and Venallt.

Until the middle of the 19th century wrought iron, produced from pig by the puddling process, was the dominant form, but in 1856 Bessemer first publicised his steel-making process and this marked the start of the conversion to steel. In the Swansea area it was the open-hearth process rather than the Bessemer process that found favour. It was invented by William Siemens who first used it at his Landore steelworks at Swansea in 1869.

37   LANDORE SIEMENS STEELWORKS
☆   **N I PS**   [SS 6679 9607]   ☐

William Siemens (1823-1883), a German by birth, came to Britain in 1843 and subsequently developed the open-hearth method of steel production. In 1869 he built the Landore Siemens Steelworks on a site to the south-west of Landore viaduct [127]. In 1871 this was expanded to the north-east on the opposite side of the river, and by 1873 it was one of the four largest steelworks in the world. Siemens steel was used to construct warships at Pembroke Dock and the Forth Bridge. The works closed in

1888 but Landore Foundry (still in operation) retains the charging-bank wall of the blast furnaces and the converted blast-engine house.

38   CLYDACH FOUNDRY
★   **N I S**   [SN 6871 0086]   ☐

This site was the first of many to use surplus lock-water from the Swansea Canal to provide water power. A waterwheel, built in 1829, provided the blast for the foundry of John Strick. Production ceased during World War II. The wheel house survives and latterly housed a water turbine, supplied by Gilkes of Kendal in either 1900 or 1917. It was taken out in the early

1980s and now lies in the yard. A timber foundry crane also survives as an integral part of the structure of the works. Pig was brought down the Swansea Canal from Ynysgedwyn to make castings at the foundry.

39 CLYDACH UPPER FORGE
★★   N I   [SN 6869 0197]   ☐

Possibly built in the late 18th century by Richard Parsons, the ironmaster of Ynysgedwyn [42] to convert pig-iron produced there. Still visible are the huge reservoir, now full of colliery waste, and a section of the dam where the Lower Clydach River runs through it. The river in fact now passes through the wheelpit and directly onto the floor of the forge. A complex of anvil and hammer beds can also be seen in the shallow river bed, and also the circular tailrace culvert. The thick squat masonry dam is very similar to that at Llandyfan [50]. The Upper Forge fell out of use at some date after 1866.

40 LLIW FORGE
☆   [SN 6060 0085]   ☐

Established c1740 in close association with Llandyfan [50]. In the latter part of the 19th century it was developed from a simple forge into a more ambitious engineering business which supplied machinery to the local tinplate industry. It probably closed at some date in the 1920s but remained intact until it was stripped for scrap in the last war. Cottages attached to the forge remain and are still occupied.

41 YSTALYFERA IRON AND TINPLATE WORKS
★   N I   [SN 7645 0830]   ☐

Built 1838 and claimed by the 1850s to be the largest tinplate works in the world. Its bank of 11 blast-furnaces was second only to that at Dowlais. All the anthracite ironworks went into decline from the 1860s and by 1864 Ystalyfera had only six furnaces in blast. The owner, James Palmer Budd, struggled to keep the works going, both for the sake of the workforce and out of personal pride, but the few remaining furnaces were blown out in 1885. There was also a 16-mill tinplate works, where production continued until after World War II. The buildings were demolished in 1946.

Modern single-storey factories now occupy most of the site which lies to the south of the village of Ystalyfera. To their west is the large stone wall of the furnace charging-bank. Much of the works site was reclaimed from the floodprone valley of the Tawe and impressive ironslag embankments contain its course to the south of the site. Nearly all the rows of houses which line the hillside between Godre'r-graig and Ystalyfera were built for workers in this business.

42 YNYSGEDWYN IRONWORKS
★   N I PS   [SN 7836 0921]   ☐

A single charcoal furnace is believed to have been built here in 1612. In 1823 George Crane, of Bromsgrove in Worcestershire, acquired the

42 Print of Ynysgedwyn Ironworks, c1820.   National Library of Wales

works and the site underwent a rapid programme of enlargement and experimentation. This really took off when Crane and David Thomas applied Neilson's hot-blast method which allowed the local anthracite to be used successfully for iron smelting for the first time. As a result of this no fewer than 36 iron furnaces were built in the anthracite coalfield in the succeeding years. At Ynysgedwyn itself there were seven furnaces at the time of Crane's death in 1846 but his successors did little to develop his works. In 1866 the remaining six furnaces were largely demolished and a new charging-bank constructed to feed two new circular metal-clad furnaces. Both of these were out of blast by 1869. In 1872 a forge or mill with monumental arches of yellow brick and a dated chimney stack were added. This project was never completed but the walls still stand. The furnace remains that were of real historical significance were cleared in 1978. Bollards cast at Ynysgedwyn survive at Swansea [62] and at Bristol. Many of the workers were housed in the nearby College Row [159] and in Gough Buildings.

43 ABERCRAVE IRONWORKS
★   N I S   [SN 8099 1260]   ☐

The first ironworks to be built expressly to use anthracite as fuel. It worked after a fashion between 1824 and 1829 but was disused until it re-opened in 1855-61. It was built by the local colliery and quarry owner, Daniel Harper, on a section of the main Swansea Canal feeder that had originally been intended to form part of the navigable canal, including two locks. This fall of water allowed a large waterwheel to be built to provide the furnace-blast. The base of a circular stone-built furnace remains, seated on a cast-iron curb with an iron 'bear' (a solidified gob of iron) dumped nearby. The stone retaining wall of the charging bank gives an idea of the original height of the furnace. The lines of a tramroad and the diverted canal leat can be traced towards Abercrave village. The buried wheel pit was excavated by the Royal Commission and the size of the wheel, 35' (10·6m) diameter, was determined by the scars which it had left on the side of the pit. The large white houses in the terrace leading from the ironworks towards the village were originally a house and the truck-shop of the ironworks.

44 BRITON FERRY IRONWORKS
☆   [SS 7345 9380]   ☐

18 An ironworks was established on the banks of the River Neath in the 1840s. It was reconstructed in the 1890s, still as an iron smelter and closed in 1958. The site has been completely cleared with the exception of the blast engine house of c1910 which housed a Richardson, Westgarth quarter crank blowing engine. It is constructed of concrete blocks in a steel frame, and has two rows of classical-style fenestration, now filled in.

45 NEATH ABBEY IRONWORKS
☆☆   N I PS   [SS 7380 9775]   **LS,SC**

The Fox family of Falmouth expanded from their original foundry in Cornwall into south Wales in 1792 when they took over an ironworks at Neath Abbey, conveniently located alongside the River Clydach a short distance from its confluence with the River Neath. The original family business passed into the control of Joseph Tregelles Price (1786-1854) in 1817, the greatest of the Neath Abbey ironmasters under whom the works gained a reputation for high quality engineering products, including stationary engines, locomotives, steamships and gas plant. After the death of Price stagnation set in and the works finally closed in 1885. Like many leading industrialists, both the Fox and Price families were Quakers.

Remaining features include two superb furnaces (1793) built against a rock face for ease of charging; Ty Mawr, the ironmaster's house (1801); and the engine manufactory. Higher up the Clydach valley is a former forge with a notable roof cast at the works in 1825 ☆ N I S [SS 7377 9803] **LS**. The building was later used as a woollen mill (machinery now in Swansea Maritime & Industrial Museum) and is now used by an overall manufacturer. Above that again the river was dammed to ensure a reserve of water and a strong steady flow to the ironworks. The present large masonry dam ★ I [SS 7392 9877] **SC**, which carries a public road, dates from c1840.

46 MELINCWRT FURNACE
☆☆   N   [SN 8243 0185]   ☐

The rather confused remains of an 18th century furnace at a spectacular location beside a high waterfall. It was started in 1708, converted from charcoal to coke, c1795, and closed in 1808. The pig which it produced was taken to the Dylais Forge at Aberdulais for conversion into wrought iron. The site is difficult to interpret but it is known to have had a blast furnace powered by a waterwheel, an 'air-furnace', a finery and foundry, and ancillary buildings. Water to power the wheel was not derived from the waterfall but was conveyed from higher up the Neath valley by a series of leats.

47 VENALLT IRONWORKS
★★   N I   [SN 8644 0499]   **SC**

Constructed c1839-42 to smelt iron with anthracite by means of the hot-blast process then

45 Neath Abbey Ironworks: the Dam.

recently introduced at Ynysgedwyn [42]. The furnaces were out of use by 1854 and the site was re-used for a patent fuel works which led to their demolition. Surviving features include the blowing-house engine, part of the cast-house, the base of the stack and the furnace walls. The office building is now a farmhouse. An interesting survival is a furnace bear with two water-cooled tuyeres embedded in it.

48   BANWEN IRONWORKS
☆☆   N I PS   [SN 8678 1042]   **SC**

The most complete example of an ironworks to survive on the anthracite coalfield, which it owes to its early failure. It was built by London speculators in 1845-48 during the years of the railway mania and may only have produced some 80 tons of pig iron. It stands as a monument to a financial scandal that extracted money from many unsuspecting shareholders. The cowhouse next to Tonypurddyn Farm was the carpenter's shop and smithy for the works, and a pond to supply condensing and boiler water to the blast engine remains to the rear. A small stone hut in a nearby field was a railway weighbridge house and the weighbridge itself survives intact although buried. To the south are the foundations and ruins of Tai-Garreg,

stone houses which housed the workers. Between the farm and the River Pyrddin is a huge masonry charging bank with two substantially intact furnaces and a crumbling blast-engine house.

49   RABY'S FURNACE, LLANELLI
☆☆   N I   [SN 5039 0158]   □

A furnace was built in Cwmddyche in 1793. The original owners were bought out in 1796 by Alexander Raby, a figure of great importance in the industrial development of Llanelli. His main interest was in the iron industry but he also worked coal, both for his ironworks and for shipping. He built the Carmarthenshire Dock [71] and was closely involved with the Carmarthenshire Railway [150], a branch of which brought raw materials into the furnace.

After taking over the ironworks, Raby enlarged it and added a second furnace, c1800. Ironmaking lasted until about 1815 when both furnaces were blown out. One of them was later demolished but the other remains in a heavily wooded cwm. On the opposite side of the B4309 road the ruins of Raby's house still remain although in a dangerous state. The modern name for this part of Llanelli, Furnace, derives from Raby's furnaces.

50 LLANDYFAN FORGE
☆☆ [SN 6588 1696] ☐

A charcoal forge which was in existence by 1669 and survived until 1807. It was located in well wooded country close to the River Loughor from which water was derived to power the blast. The ruins of the forge itself are obscured by thick undergrowth and marshes, but the 200' (60m) long dam wall with two waterwheel openings in it can be identified. Some 300 yds (275m) downstream are remains of the pond and wheel pit of Llandyfan New Forge (c1780-1807) at SN 6563 1683. The other ruins on this site are of a late 19th century woollen mill.

For further sites at which there are remains of the iron and steel industries see entries 55 and 168.

TINPLATE

Tinplate manufacture was—and to some extent still is—a highly characteristic industry of the Swansea region. In 1913 four out of every five tinplate workers in the United Kingdom lived within a 20-mile radius of Swansea, and towns such as Llanelli, Morriston, Pontardulais, Briton Ferry, Gorseinon and Pontardawe were almost entirely dependent on the tinplate trade for their economy. Tinplate—iron or steel bar rolled into sheets and coated with a wash of tin—was first manufactured in Britain at Pontypool in the 17th century. During the next hundred years a number of works were established in south Wales and elsewhere. The first works in Glamorgan was at Ynyspenllwch (1747), near Clydach in the Swansea valley, but it was not until 1845 that the industry arrived in Swansea. There were a number of reasons for this area proving so suitable for it. As well as convenient supplies of coal and water, there was a reserve of metallurgical skill in the area built up through the smelting industries. Also significant is the fact that from the middle of the 19th century the ferrous industries were increasingly dependent on imported ores which resulted in the location of steelworks, and hence tinplate manufacture on the coastal strip.

Unlike the copper industry, where the capital came from outside the region, tinplate was to a large degree financed and managed locally. The larger concerns were integrated businesses comprising both steelworks and tinning shops, but there were also many smaller works which bought their steel from outside suppliers.

It was the American market which really set up the tinplate trade. It proved a cheap material for all the domestic utensils a homesteading family would need, including a roof for their house. It formed the cans in which the meat-processing factories of Chicago packed their bully beef and the drums in which the nascent oil industry transported petroleum. But the Swansea tinplate industry was too successful for its own good and in 1891 President McKinley imposed a tariff on imported tinplate. Swansea was badly hit: many of the workers migrated to the United States and contributed to the success of its growing industry, while those who remained had to face a period of depression until the trade could re-establish itself with new outlets on the continent and in the Far East.

Tinplate as traditionally manufactured was a labour-intensive industry. It had a low rate of worker-productivity compared to the more highly automated American industry. After World War II, therefore, to ensure that Wales retained a tinplate industry, a process of rationalisation was devised. This saw the replacement of the old handmill by the modern automated strip mills at Trostre (Llanelli) and Felindre (north of Swansea). Both are still in production, tinning steel coil manufactured at Port Talbot.

Several tinplate works still survive, although gutted and reused for new purposes. Few of them can claim to have any architectural merit—they are large, gaunt, unadorned and four-square. Nevertheless they have a certain brooding presence, and there is more than a dash of chapel in their architecture, which shows just how typically Welsh the tinplate trade was—the religion of the gwerin echoed in the industry of the gwerin.

51 BEAUFORT TINPLATE WORKS, SWANSEA
☆   **N I PS**   [SS 6710 9705]   **LS**

Established in 1860 on the site of the Fforest copper mills by John Jones Jenkins (later Lord Glantawe), a figure of major importance in the local tinplate trade. Sold by Jenkins in 1877, and closed in 1946. Two main blocks survive with various ancillary buildings. The more easterly of these, the annealing house, is built of stone and can be dated to 1874 by the cast copper slag keystone of the main access arch. Remains of tinning bays are believed to survive within the second block. The site is now used as a headquarters of the local YTS and casual access is not possible.

52 PONTARDAWE TINPLATE WORKS
☆   **N**   [SN 7220 0364]   ☐

Founded 1843 by William Parsons as the Primrose Forge and Tinplate Works. It was powered by water from the Pontardawe tucking mill and from the adjacent Swansea Canal. Open-hearth furnaces were added later to make it an integrated steel and tinplate works, and under later owners, the Gilbertsons, Pontardawe became virtually a one-company town until the closure of the works in 1962. Most of the site has now been cleared except for one large late 19th century building still in industrial use. It is one of the very few structures to retain its tinning bays and a floor of cast-iron slabs.

53 GWALIA TINPLATE WORKS, BRITON FERRY
☆   [SS 7339 9377]   ☐

Erected 1892 to use steel produced by the Briton Ferry ironworks [44]. Acquired by the Briton Ferry Steel Co in 1937 and closed in 1953. The building has been truncated and only five of the stone-built bays survive. The chimney stacks have been dismantled and the building re-roofed but brick flues in the walls mark the original hearths. A brick-built extension to the west carries a plaque dated 1898-1899. On an adjacent site to the north was the Villiers tinplate works (1888) of which some modern brick-built buildings remain.

54 ABERDULAIS FALLS
★★   **N I PS**   [SS 7717 9950]   **NT**

A highly attractive site beside a waterfall in a wooded gorge which endeared itself to artists of the 18th and 19th centuries. It is now the property of the National Trust who have recently carried out a programme of archaeological work.
Aberdulais Falls was probably the site of 16th century copper smelting, although there are no tangible remains. It subsequently was used for a corn mill, ironworking and tinplate manufacture. The visible remains belong to the tinplate period and were presumably the work of the Aberdulais Tinplate Co, founded 1830. This company had a split site, with the upper works at the Falls and the lower works approximately 400 yds to the south, on the left bank of the River Neath, on the site of the former Dylais Forge. The upper works is thought to have been abandoned c1890 although the lower works continued to produce tinplate until 1939.

The main features are:-
– weir, header tank and wheel pit, c1840.
– the 'Bastion', a ponderous and imposing, yet foundationless structure of dressed stone, the function of which is not clear, c1840.
– a 60' (18·3m) high stack, probably also c1840.
– foundations of rolls and furnaces.
– masonry bridge over the River Dulais which carried a tramway to a dock beside the Tennant Canal. Only one of the original two arches remains.

55 SOUTH WALES STEEL AND TINPLATE WORKS, LLANELLI
☆   [SS 5082 9858]   ☐

Started as a tinplate works in 1872 with capital from Birmingham. In 1879-80 two open-hearth furnaces were erected, making this the first integrated steel and tinplate works in Llanelli. Purchased by Richard Thomas & Co in 1899 who added eight new mills to the existing ten in 1911 and introduced the continuous process of tinplate production. A plaque on the wall of the building facing the New Dock [73] records this extension which made the South Wales works one of the largest in the region. It remained part of the Richard Thomas empire (subsequently RTB and then Steel Company of Wales) until closure in 1958.

56 OLD CASTLE TINPLATE WORKS LLANELLI
☆   **N I**   [SN 4990 0020]   ☐

Established in 1866-67 and expanded early in the 20th century. Tinplate manufacture ceased in 1957 and the works are now used for other purposes. Very plain, unadorned construction, mostly of stone.

57 ABERLASH TINPLATE WORKS AMMANFORD
☆   [SN 6230 1305]   ☐

Erected 1889 and closed in 1908. The building is now used by car dismantlers and is ivy-grown and deteriorating. It is constructed of stone with brick surrounds to the doors and windows. The southern facade comprises a central arched doorway, large enough to give railway access, with five lancet windows above it, stepped in height so as to match the gable end. The roof is original and in an apparently dangerous state. One of the chimney stacks from the tinning bays survives.

58   ASHBURNHAM TINPLATE WORKS, BURRY PORT
☆   **N I**   [SN 4416 0040]   □

Founded 1890 and in production until 1953. Architecturally very plain, constructed of stone relieved with brickwork. Externally unaltered and now in use by a pickle-packing firm.

59   KIDWELLY TINPLATE WORKS AND MUSEUM
★★   **N I S**   [SN 4212 0040]   □

Tinplate manufacture started at Kidwelly in 1737 on the site of an earlier forge and continued until 1941. It was the second tinplate works to be erected in Britain. The works area has now been redeveloped as an industrial museum. Among the relics of the tinplate works are a water-powered mill, a plaque of 1801 recording a rebuilding of that year, and a fine steam engine which powered the cold rolls. In addition a number of artefacts have been assembled to illustrate the industrial history of the Kidwelly/Llanelli area. These include items used in the traditional method of tinplate manufacture and coal-mining equipment. Among the latter are pithead gear and the winding engine from Morlais Colliery. The museum also houses a steam crane and two locomotives.

For other sites at which tinplate was manufactured, see entries 41 and 42.

59 Rolls Engine at Kidwelly Tinplate Works.

RIVER AND PORT FACILITIES

The river navigations of south-west Wales were not ambitious engineering works which extended navigation far inland by means of cuts and locks. They were, however crucial to the industrial development of the region. It was the penetration of the coalfield by tidal rivers that led to the early development of a sea-borne export trade in coal through the ports of Swansea, Neath and Llanelli.
Initially the techniques used for handling these shipments of coal were primitive: the vessels simply tied up alongside the river edge or sailed up a narrow pill at high water and settled on their bottoms as the tide went out. Once loaded, they waited for the next tide to float them off. The first improvements came in the 16th century when quays started to be built along the waterfront in Swansea or on the sands of Swansea Bay between high and low water, and in the 17th century the first docks—tidal, of course—started to be constructed. It was also at this period that the first attempts were made to extend river navigation inland: a pound lock was built at Aberdulais in 1699, followed by a second one in the 1740s, to make the River Neath navigable to Ynys-y-gerwyn tinplate works, eight miles from the coast.
The export of coal started to be supplemented during the 17th century by the import of copper ore and lead ore. This trade tended to be handled at river wharves along the Neath and Tawe rivers. Indeed, one of the factors that attracted the nonferrous smelting industry to this region was the existence of rivers navigable for several miles inland, and with suitable sites alongside them for constructing smelting houses.
It was the growth of the coal trade that produced many major schemes for harbour improvement. The first floating dock in south Wales was constructed by Humphrey Mackworth at Neath in the early 18th century. The first public floating dock was opened in 1834 at Llanelli, but it was not until 1852 that Swansea, whose harbour had been managed by a public Harbour Trust since 1791, had its first floating dock.
Until well into the 19th century Swansea was the main coal shipping port of south Wales simply because the western half of the coalfield was worked more intensively than the east. But with the improvements in inland transport provided by canals and even more by locomotive railways, the eastern coalfield was opened up and the western ports were overtaken by Newport, Cardiff and later Barry.
The three main ports of the western coalfield had traditionally been Swansea, Neath and Llanelli, but the increasing economic activity of the 19th century led to the development of several smaller ports. Aberafan, always handicapped by difficult approaches, was developed in the 1830s as Port Talbot, and other new ports were created at Pembrey, Burry Port and Briton Ferry. In contrast. the ancient port of Neath went into decline because it continued to rely on riverside wharves and so lost out to the floating dock at Briton Ferry. in the 1880s an attempt was made to modernise the port but without success.
Much of the dock development in the area was undertaken in close conjunction with railway promotion. Sometimes the same company developed both docks and railway as at Llanelli and Port Talbot (and, *par excellence*, although outside our area, Barry). At Burry Port the dock company and a canal company combined to form a railway. In other cases close collaboration existed between the promoters of the railway and the dock, as between the South Wales Mineral Railway and Briton Ferry. At Swansea the same sort of symbiotic relationship existed: the Rhondda & Swansea Bay was promoted by an independent group to attract traffic from the Rhondda to the town's new Prince of Wales Dock, and both the GWR and the Midland Railway planned new lines to link the anthracite coalfield with the docks—even if they did not reach completion—at the same time as the Harbour Trust was building the King's Dock, also for the anthracite export trade.
The south Wales coalfield reached its highest ever output in 1913, much of the yield going for export. Thereafter the coal trade contracted and this is reflected in the history of the docks. One by one coal traffic ceased, so that eventually only Swansea was handling this traffic, and it is quite

likely that by the end of 1987 coal shipping will have ceased here too, leaving Barry as the only coal port in south Wales. There are now only two registered ports in the region, Swansea and Port Talbot, although commercial shipping is also carried on at wharves along the River Neath. In several cases leisure sailing has taken over from commercial traffic and former coal docks now act as yachting marinas.

SWANSEA DOCKS N I   LS (part)

By Elizabethan times loading stages or quays were being built along the banks of the River Tawe. The line of the original waterfront is now represented by The Strand. The first dock appears to have been the 'towne dock' of 1624 and by the beginning of the 18th century warehouses were in existence along The Strand. Also by this date the main coal producers had all acquired their own 'coale places' on the banks of the river. The first serious attempt at improving the facilities was made in 1768, and in 1791 the first Harbour Act was passed which resulted in the creation of the Swansea Harbour Trust, the body which administered the port until 1922. They built the East Pier c1800 of which part remains [SS 6640 9257]. An important development was the creation of Port Tennant, a tidal harbour on the eastern side of the river, in 1824 as the terminus of the Tennant Canal [95–98]. The first floating dock, the North Dock, was opened in 1852. This was formed by diverting the river into its present course, the New Cut (1842-45) and making its original course into the dock. The South Dock followed in 1859, built with an eye to the export trade in coal from Aberdare. Both these docks were on the western side of the river. In 1881 the Prince of Wales Dock opened, the first of the modern complex of docks on the eastern side. It was built largely on the site of Port Tennnant. The King's Dock opened in 1909 for the export of anthracite, and the last of the docks, the Queen's Dock, opened in 1920. This was for the import of crude oil for the refinery of the Anglo-Persian Oil Company (now BP) at Llandarcy.

The North Dock closed in 1928 although the basin remained in use until the 1960s. Commercial shipping in the South Dock ceased in 1971 and it has now been developed as a marina. Commercial traffic is confined to the eastern docks.

Remaining features include:-

60   HARBOUR TRUST OFFICES
★   [SS 6599 9287]   **LS** (II☆)

Opened 1903; built of brick and Portland stone in the Baroque style.

61   SOUTH DOCK
★   [SS 6575 9250]   WAREHOUSE PART: **LS**

Now a yachting marina with extensive new development on all sides. The former Coast Lines warehouse houses the city's Maritime & Industrial Museum.

62   NORTH DOCK
★   [SS 6587 9345]   ☐

The site of the filled-in dock has recently been redeveloped for retailing. The entrance to the half-tide basin can be seen behind Sainsbury's, while in front of it various bollards have been repositioned. Two further bollards, one inscribed 'Yniscedwyn Ironworks' [42] can be seen to the north of the New Cut bridge.

63   PUMP HOUSE
★   [SS 6601 9264]   **LS**

Hydraulic power house dated 1900, located between the inner and outer basins of the South Dock. The bridge which it formerly operated remains alongside but reduced to a mere half of its original length.

64   VICTORIA QUAY
★   [SS 6567 9257]   **LS**

Rubble revetment walls with six bastions at irregular intervals which gave access to coaling stages.

65   PILOT HOUSE
★   [SS 6640 9256]   **LS**

Late 19th century building at the pierhead on the western side of the river.

66   *LADY QUIRK*
★   [SS 6614 9340]   ☐

The skeleton of a two-masted barque can be seen in the mud on the eastern bank of the river above the New Cut bridge. She was engaged in the coal trade until about 1910 when she was tied up and has since slowly rotted away.

There was formerly an intensive use of the River Tawe by sea-going vessels engaged in the ore and coal trades. By the beginning of the 19th century the three-mile dredged and navigable length of the river had an almost continuous line of stone-built quays and tidal dock basins along it. Good examples survive at Morfa [28], White Rock [29] and Upper Bank [30]. Historically, the most important is probably:-

67   **LANDORE QUAY**
★   **N I**   [SS 6627 9588]   □

The oldest surviving quay, built 1772-74 as the 'new quay' for the coal and copper magnate, John Morris I by William Edwards who carried out several commissions in the valley [153, 157]. Morris regarded the quay as a major innovation in constructional technique, possibly because it was the first masonry quay built locally using hydraulic mortar from Aberafan [cf 81]. Earlier quays may all have been constructed of timber-revetted earth.
For Penclawdd dock, see entry 86.

68   **BRITON FERRY DOCK**
★   **N PS**   [SS 7370 9360]   □

Opened in 1861 and constructed to designs by Brunel. It comprises an inner floating dock and an outer tidal basin with an overall area of about 18½ acres. Entry to the dock was through a single buoyant lock gate. It closed in 1959 and is now largely silted up although the walls are intact. Still to be seen are the partially dismantled lock gate and, on the eastern side of the dock, the square tower of the engine house which generated hydraulic power for its operation.

69   **NEATH FLOATING HARBOUR**
★   **N I PS**   [SS 7345 9620]   □

In 1878 the Neath Harbour Commissioners obtained an Act of Parliament to build a large floating dock in a loop of the River Neath between Neath Abbey and Briton Ferry. Uncertain trade prospects and the effect of a storm in 1885 halted the scheme but only after very substantial works had been completed. Unfortunately the extension of the local authority's rubbish tip has resulted in the dynamiting of the hydraulic power house in 1988, and the substantial mile-long by-pass canal for the tidal river is now being filled in.
One side of the large granite-built lower ship lock remains [SS 7294 9515].
See also entries 96-97.

70   **PORT TALBOT DOCKS**
☆   [SS 7648 8901]   □

Aberavon Harbour Co was formed in 1834 to build a floating dock to serve the iron and copper industries of Cwmavon. It was opened in 1839 under the designation 'Port Talbot' in honour of the local landowning family. By 1890 it was almost moribund and a new company was formed to revive it as an outlet for coal from the western central valleys. This company, the Port Talbot Railway & Dock Co [143-45] was

68 Briton Ferry Dock, c1865.   *Royal Institution of South Wales.*

formed in 1894 and by 1898 had built three new railways and made a start on improving the docks. Dock equipment was hydraulically operated and to supply water for this purpose a reservoir was built at Brombil [SS 798 879]. They were also an early user of reinforced concrete for wharf construction.

Coal ceased to be shipped in 1962, but by then the import of iron ore for the steelworks had become the main item of trade. In 1972 the old docks were replaced by a new deepwater tidal harbour specifically intended for unloading large iron ore carriers.

LLANELLI DOCKS

Coal was certainly being shipped at Llanelli in the 16th century if not before, but the earliest known coal bank to have been specially built for the purpose is believed to have been at Spitty Bank [SS 5604 9813]. The first tidal dock, Pemberton's Dock [SS 5010 9930] was developed from a tidal wharf at some date between 1794 and 1804. It was abandoned by the middle of the 19th century and no traces remain. Subsequently a number of other docks have been built in Llanelli. They are all now closed to commercial chipping but residual use on a limited scale still takes place.

71 CARMARTHENSHIRE DOCK
★   N I   [SS 4991 9950]   ☐

Llanelli's second dock was developed from a shipping place constructed by Alexander Raby in 1799. Raby sold it to the newly formed Carmarthenshire Railway in 1802 and this company then proceeded to enlarge and improve it. The dock gradually declined in importance with the construction of floating docks in the town.

72 COPPERHOUSE DOCK
☆   [SS 5053 9905]   ☐

Built 1804-05 as a tidal dock to serve the copperworks of R J Nevill. Reconstructed as a floating dock in 1824-25. It remained in use until 1949 and has since been filled in, although part of the entry channel remains.

73 NEW DOCK
★   [SS 5095 9873]   ☐

Constructed by the Llanelly Railway & Dock Co [146–49] as an outlet for coal from the pits of the Llangennech Coal Co with which the dock company had close relations. Opened in 1834 as the first *public* floating dock in Wales. Fell into disuse after the war and has now been filled in. The scouring reservoir and the ship channel remain.

74 NORTH DOCK
★★   [SS 4980 9950]   ☐

Originally a scouring basin built in 1859-62 on the recommendation of Brunel to improve the approaches to Llanelly harbour. Converted into a dock in 1898-1903 in an attempt to attract anthracite traffic. The hydraulic power house at the north-west corner is a pleasant feature.

75 BURRY PORT HARBOUR
★★   N I   [SN 4454 0028]   ☐

Pembrey New Harbour, or Burry Port, was opened in 1832. It was linked to the anthracite mines of the Gwendraeth Fawr by the Kidwelly & Llanelly Canal [100–04] and to the Carmarthenshire Dock at Llanelli [71] by tramroad. The original works consisted of a tidal channel scoured out of the sand dunes by the River Derwydd, along which was built a stone wharf.

71 Carmarthenshire Dock, Llanelli.

A scouring reservoir kept the channels free of silt. By 1840 a small floating dock, the East Dock, had been added alongside the scouring reservoir, which in turn was converted into the West Dock in 1888. Commercial shipping ceased soon after the war but the harbour remains in use for private yachting.

One of the most interesting relics is a string of iron canal tub boats on the eastern face of the east breakwater: it is believed that they were siezed from Carway Colliery in 1870 to recover a debt owing to the railway company. They are about 20′ (6·1m) long with a 5′ (1·5m) beam and flat sterns. They were probably pulled along the canal in trains.

To the north of the harbour, close to the railway station, is the former harbour hotel, dated 1841 by a plaque on the rear wall. It was subsequently much extended and now functions as a club.

76		**PEMBREY OLD HARBUR**	
★	**N I**	[SN 4370 0015]	☐

Built in 1819 to serve pits at Gwscwm and Pembrey to the north. It was a natural creek adapted for coal shipping. Additional traffic was generated with the completion of Pembrey Canal in 1824 which provided a link from the harbour to the Kidwelly & Llanelly Canal [100–04]. Both harbour and canal were superseded when Pembrey New Harbour (i.e. Burry Port) [75] was reached by the Kidwelly Canal in 1837.

77		**KIDWELLY QUAY**	
★	**N I S**	[SN 3977 0638]	☐

Built c1768 by Thomas Kymer to ship anthracite carried on his canal [99]. Trade was good until 1800 but then Kidwelly started to go into decline as a harbour owing to the silting of the river estuary. Attempts were made to revive it, but the opening of Burry Port [75] ensured that nothing came of this. Occasional shipments of coal continued until final closure in 1934. The quay has recently been tidied up by a government-funded scheme.

LIGHTHOUSES

Closely associated with harbour improvement is the provision of lighthouses to mark the approaches more clearly. Two interesting examples exist on the coast near Swansea.

78		**MUMBLES LIGHTHOUSE**	
★	**N I S**	[SS 6349 8717]	**LS**

Constructed by Swansea Harbour Trust in 1791-95. Built of stone and consists of two octagonal stages, the higher one of smaller diameter than the lower. Firebaskets were set on each stage so as to show two lights one above the other. These constricted arrangements may well have hastened the conversion of the lighthouse to oil-burning in 1798-99. The architect was William Jernegan, a local man who carried out many commissions for the gentry of Swansea in this period. The structure remains substantially unaltered although of course the light is now electric.

79		**WHITFORD POINT LIGHTHOUSE**	
★★	**N I S**	[SS 4438 9728]	☐

A cast-iron tower on the north coast of Gower built by the Llanelly Harbour Commissioners in 1865 to mark the approach channel to the port. It is the only major cast-iron wave-swept tower in British waters. Constructed of seven rings of cast-iron plates bolted together by means of external flanges, strengthened with massive wrought-iron bands. There is an attractive bracketed balcony with gothic balusters. It has long been disused and access to the interior is impossible. Good views of it may be had from Pembrey and Burry Port.

79 Whitford Point Lighthouse.   *Picture: Dylan Roberts*

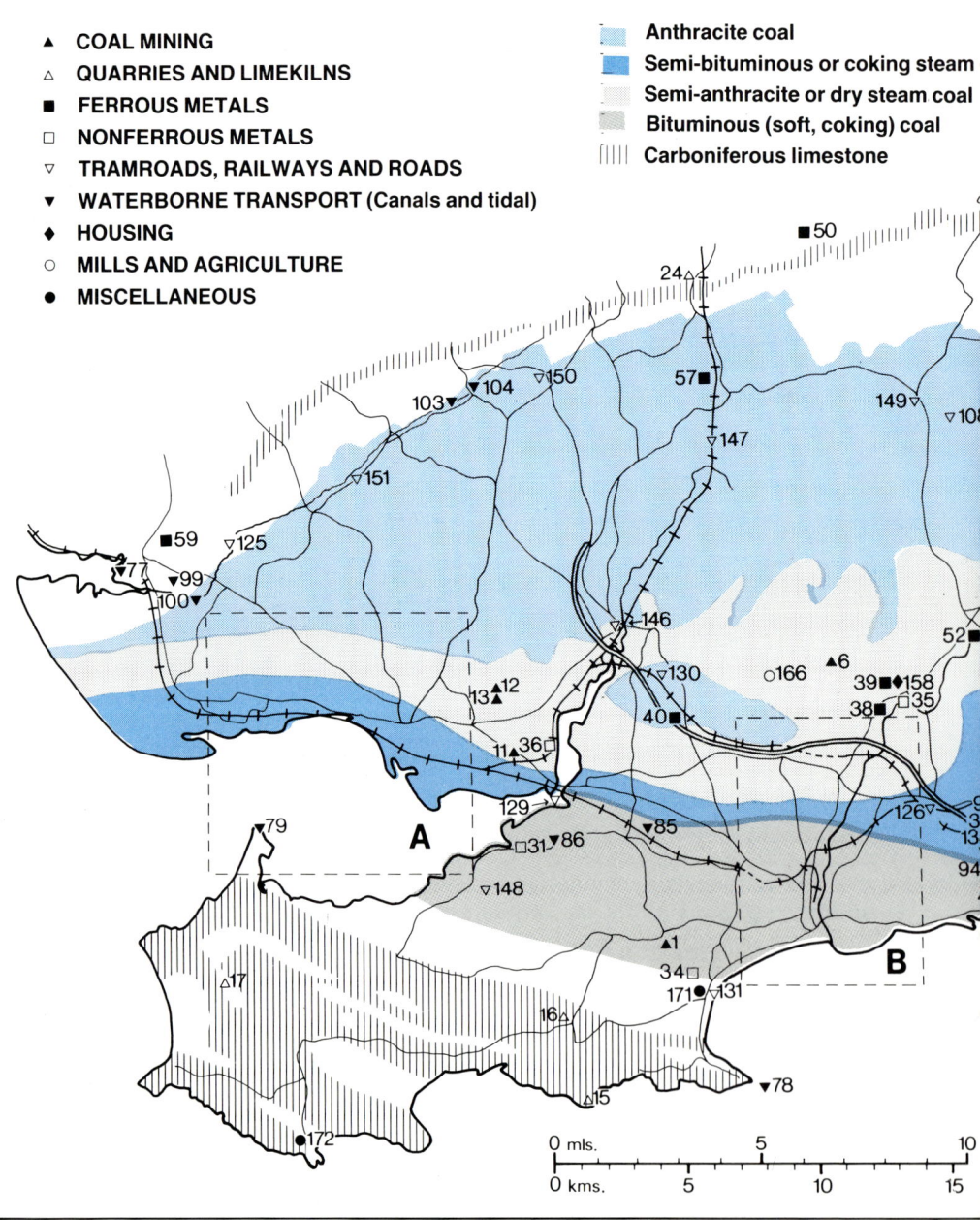

Site Plan

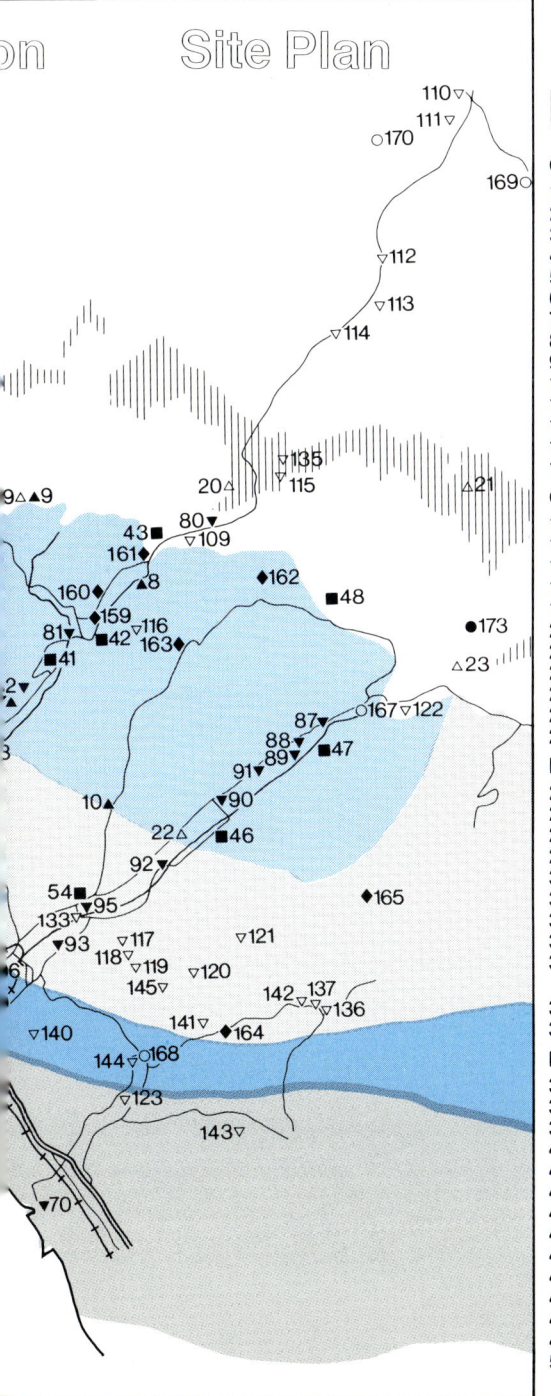

Key

COAL MINING

1	Rhydydefaid Boat Level	★
2	Pwll Mawr (Great Pit)	★
3	Gwernllwynchwith Engine House	★★
4	Scott's Pit, Llansamlet	★★
5	Garth Pit	★
6	Felindre Colliery Engine House	★★
7	Crimea Colliery	★★
8	Ynysgedwyn Colliery Ventilation Shaft	★
9	Henllys Vale Colliery	
10	Cefn Coed Winding Engine	★★
11	Genwen Engine House	★★
12	Penprys Pit	★★
13	St David's Pit	★
14	Pwll Colliery Engine House	★

QUARRIES AND LIMEKILNS

15	Pwll-Du Quarry	★
16	Vennaway Limekiln	★
17	Llangenydd Limekiln	★
18	Hafod Limekiln	★
19	Henllys Vale Limekilns	★
20	Cribarth Mountain Quarries	★★
21	Carnau Gwynion Limekilns	★★
22	Ty Bont Limekiln	
23	Dinas Silica Mine	☆☆
24	Llandybie Limekilns	☆☆
25	Foel Fawr Limekilns	★

NONFERROUS METALS

26	Llangyvelach Copperworks	★
27	Hafod Copperworks	☆☆
28	Morfa Copperworks	☆☆
29	White Rock Copperworks	★★
30	Upper Bank Copperworks	★
31	Penclawdd Copperworks	★
32	Crown Copperworks	☆
33	Pembrey Copperworks	☆
34	Clyne Wood Arsenic and Copperworks	★★
35	Mond Nickel Works, Clydach	★
36	Pencoed Leadworks	☆

FERROUS METALS

37	Landore Siemens Steelworks	☆
38	Clydach Foundry	★
39	Clydach Upper Forge	★★
40	Lliw Forge	☆
41	Ystalyfera Iron and Tinplate Works	★
42	Ynysgedwyn Ironworks	★
43	Abercrave Ironworks	★
44	Briton Ferry Ironworks	
45	Neath Abbey Ironworks	☆☆
46	Melincwrt Furnace	☆☆
47	Venallt Ironworks	★★
48	Banwen Ironworks	☆☆
49	Raby's Furnace, Llanelli	☆☆
50	Llandyfan Forge	☆☆

51 Beaufort Tinplate Works, Swansea ☆
52 Pontardawe Tinplate Works, Briton Ferry ☆
53 Gwalia Tinplate Works, Briton Ferry ☆
54 Aberdulais Falls ★★
55 South Wales Steel and Tinplate Works, Llanelli ☆
56 Old Castle Tinplate Works, Llanelli ☆
57 Aberlash Tinplate Works, Ammanford ☆
58 Ashburnham Tinplate Works, Burry Port ☆
59 Kidwelly Tinplate Works and Museum ★★

WATERBORNE TRANSPORT
60 Swansea Harbour Trust Offices ★
61 Swansea South Dock ★
62 Swansea North Dock ★
63 Swansea Pump House ★
64 Victoria Quay ★
65 Swansea Pilot House ★
66 *Lady Quirk* barque ★
67 Landore Quay ★
68 Briton Ferry Dock ★
69 Neath Floating Dock ★
70 Port Talbot Docks ☆
71 Carmarthenshire Dock, Llanelli ★
72 Copperhouse Dock, Llanelli ☆
73 New Dock, Llanelli ★
74 North Dock ★★
75 Burry Port Harbour ★★
76 Pembrey Old Harbour ★
77 Kidwelly Quay ★
78 Mumbles Lighthouse ★
79 Whitford Point Lighthouse ★★
80 Abercrave Feeder Weir ★
81 Ystalyfera Aqueduct and Weir ★★
82 Godre'r Graig Lock and Dry Dock ★
83 Ynysmeudwy Canal Docks and Tramroad ★★
84 Smith's Canal ★
85 Gowerton Lock ☆
86 Canal Dock and Scouring Basin, Penclawdd ★
87 Aberpergwm Aqueduct ★
88 Maes Gwyn Boathouse, Mill & Limekiln ★
89 Cwm Gwrach Branch Canal ★
90 Resolven Lock and Aqueduct ★★
91 Rheola Brook Aqueduct ★★
92 Ynysbwllog Aqueduct ★
93 Neath Canal Depot, Tonna ★★
94 Red Jacket Pill Lock ★★
95 Aberdulais Aqueduct and Basin ★★
96 Clydach Aqueduct, Neath Abbey ★
97 Skewen River Dock and Lock ★★
98 Skewen Cutting ★★
99 Kymer's Canal ★
100 Spudder's Bridge Aqueduct ☆☆
101 Pembrey Bridge ★
102 Stanley's Bridge ★

8 Ynysgedwyn Colliery ventilating tower.
Photograph: S K Jones.

TRAMROADS, RAILWAYS AND ROADS
103 Hirwaunissa Incline Plane ☆
104 Cwmmawr Basin ★
105 Scott's Tramroad ★
106 Mynydd Newydd Railroad Incline ★
107 Pentrefelen Railroad ★
108 Gwaun-cae-Gurwen Railway ☆
109 Yard Bridge ★
110 Sennybridge Tramroad Depot ☆☆
111 Clydach Tramroad Causeway ☆☆
112 Cnewr Tramroad Depot ☆☆
113 Grawen Tramroad Depot ☆☆
114 Bwlch Bryn-rhudd Causeway ☆
115 Penwyllt Quarry Tramroads ★★
116 Claypon's Incline and Engine-House ★★
117 Parsons' Railway Inclines ★★
118 Parsons' Summit Level ★★
119 Parsons' Incline and Engine-House ★★
120 Tonmawr Village Tramroad ★
121 Fforchdwm Viaduct ★
122 Glynneath Inclined Plane ★★
123 Bryn Railroad ★
124 Stanley's Tramroad, Pembrey ★
125 Pwll y Llygod Bridge ☆☆
126 Llansamlet flying arches ☆
127 Landore Viaduct ★★
128 Swansea (High Street) Station ★

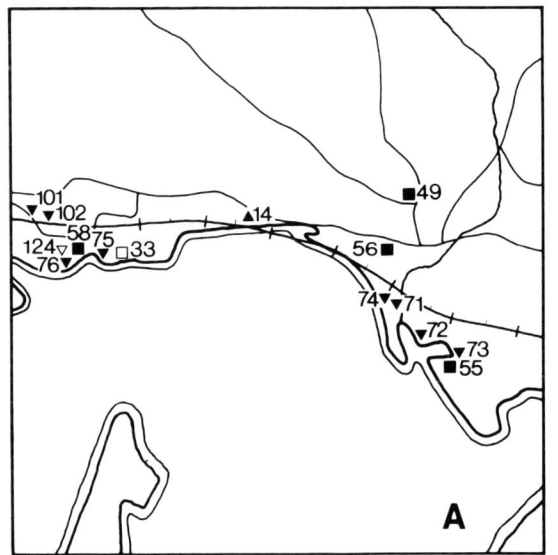
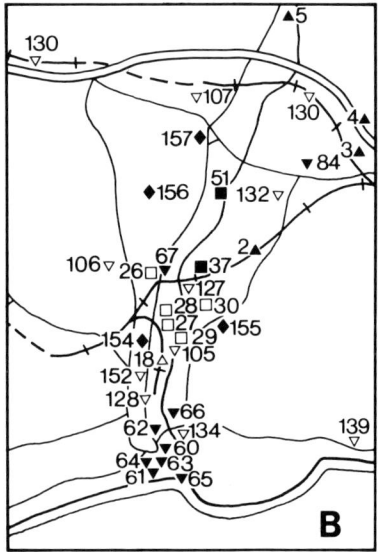

Llanelli & Burry Port   Swansea

129	Loughor Viaduct	★
130	Swansea District Line	
131	Swansea & Mumbles Railway	★
132	Swansea Vale Railway	
133	Aberdulais Viaduct	★
134	Swansea (New Cut) Bridge	★
135	Craig-y-Nos Station	★
136	Croeserw Viaduct	★
137	Cymmer Station	
138	Neath River Bridge	★★
139	Danygraig Locomotive Shed	☆
140	Ynys-y-Maerdy Incline	★
141	Gyfylchi Tunnel	★
142	Cymmer Viaduct	★
143	Cwm Cerwyn Tunnel	★
144	Pontrhydyfen Viaduct	★
145	Tonmawr Junctions	★
146	Pontardulais Tunnel	☆
147	Pantyffynnon Station	★
148	Llanmorlais Station	☆
149	Gwaun-cae-Gurwen Viaducts	★
150	Llanelly & Mynydd Mawr Railway	
151	Burry Port & Gwendraeth Valley Railway	
152	Hafod Bridge	★
153	Pontardawe Bridge	★

HOUSING

154	Vivianstown	★
155	Grenfell Town	★
156	Morris Castle	★★
157	Morriston	★★
158	Forge Row, Clydach	
159	College Row, Ynysgedwyn	★★
160	Gough's Row, Ystradgynlais	★
161	Sheperds' Cottages, Cae'r-bont	★★
162	Price's Row, Coelbren	★
163	Church Street, Seven Sisters	★
164	Cynonville Garden Village	★
165	Brick Street, Glyncorrwg	★

MILLS AND AGRICULTURE

166	Felindre Watermill	☆
167	Glynneath Woollen Mill	★
168	Bont Fawr Aqueduct, Pontrhydyfen	★★
169	Forest Lodge Model Farm	☆☆
170	Belfont Model Farm	☆

MISCELLANEOUS

171	Blackpill Pyroligneous Acid Works	☆
172	Port Eynon Salthouse	★★
173	Glynneath Gunpowder Works	★★

CANALS

The first canals in the area, built between 1696 and the 1790s, were merely short affairs to improve transport between the copper- and iron-smelting works, collieries and shipping staithes in the estuarine areas. One of these canals, at the Forest copperworks near Morriston was largely underground: it is believed to have been built in 1746 and is stated to have reached into the hillside for one mile by c1800. The remains of what appear to be a similar arrangement can be seen in the Clyne Valley Country Park in Swansea. None of these early canals were longer than four miles.

In 1791-99 the first of the longer trunk canals was built to open up the coal industry in the upper Neath valley. It was some 13 miles in length. The similar Swansea Canal of 1794-98 was 16¼ miles in length. Both assumed subsidiary roles as water-power suppliers for the ironworks and factories that were opened along them.

The second decade of the 19th century saw a further expansion in canal construction. The early Kymer's Canal in the Gwendraeth valley was extended to a length of nine miles as part of the Kidwelly & Llanelly Canal; the collieries to the west of Swansea were given access to the sea by the 3½ mile Penclawdd Canal and the Swansea and Neath estuaries were connected by George Tennant who then extended his canal from Swansea harbour to the Neath Canal at Aberdulais to complete the 8½ miles of the Tennant Canal by 1824.

SWANSEA CANAL

The last of the major canals of south Wales, it was incorporated in 1794 and fully opened in 1798. It was basically a speculative undertaking designed to open up the coal trade at the head of the valley. It was bought by the GWR in 1873 and worked by them as a competitor to the Midland Railway's Swansea Vale line. It ceased to be profitable after 1902 and was closed to all traffic in 1931. Much of the canal has now been filled in or used for road construction, but it is still in water where it passes through Clydach and Pontardawe.

80 ABERCRAVE FEEDER WEIR
★ N I S [SN 8209 1264] SC

A stone-built weir which was the main source of water for the canal. It also provided water to drive some 40 water-powered installations sited along the canal. It has been rebuilt several times as the '1842' date stone indicates.

81 YSTALYFERA AQUEDUCT AND WEIR
★★ N I S [SN 7727 0924] SC

The largest aqueduct on the Swansea Canal, built by Thomas Sheasby in 1794-98, using hydraulic mortar from Aberafan. The Swansea Canal aqueducts were probably the first in Britain to use such mortar as the waterproofing agent instead of puddling clay. It consists of three segmental arches built on top of a feeder weir, the crest of the weir being paved to prevent any scouring of the foundations. A large circular culvert through the northern end of the aqueduct carried the tailrace water from a fulling mill at Gurnos.

82 GODRE'R GRAIG LOCK AND DRY DOCK
★ N I S [SN 7582 0738] SC

Substantial remains exist of lock 17 and of a dry dock which was added to the lock platform after 1875. The lock by-pass was used to drain the dock when required for the maintenance of boats. There are also remains of a pitch boiling hearth and of the timber uprights which supported cross-pieces to carry boats under repair.

83 YNYSMEUDWY CANAL DOCKS AND TRAMROAD
★★ N I PS [SN 7377 0500] SC

Until 1837 Ynysgedwyn ironworks [42] depended on bituminous coal brought in from lower down the Swansea valley. In 1828 the owner of the works, George Crane, opened a colliery at Cwm Nant Llwyd [SN 7402 0380] as a source of supply and connected it to the Swansea Canal by a tramroad and branch canal. At the head of the incline [SN 7374 0459] which took the tramroad down to the valley floor are the stone supporting walls of a winding drum. This was designed by William Brunton and used the weight of loaded wagons descending the incline to haul wagons along the level section from the colliery to the incline head. This section includes a fine stone-revetted causeway [SN 7377 0443]. At the foot of the incline, which is carried on a stone retaining wall, the portal of another coal level, refronted in the 1860s, is visible. A timber trestle carried the tramroad across the River Tawe, and the abutments and

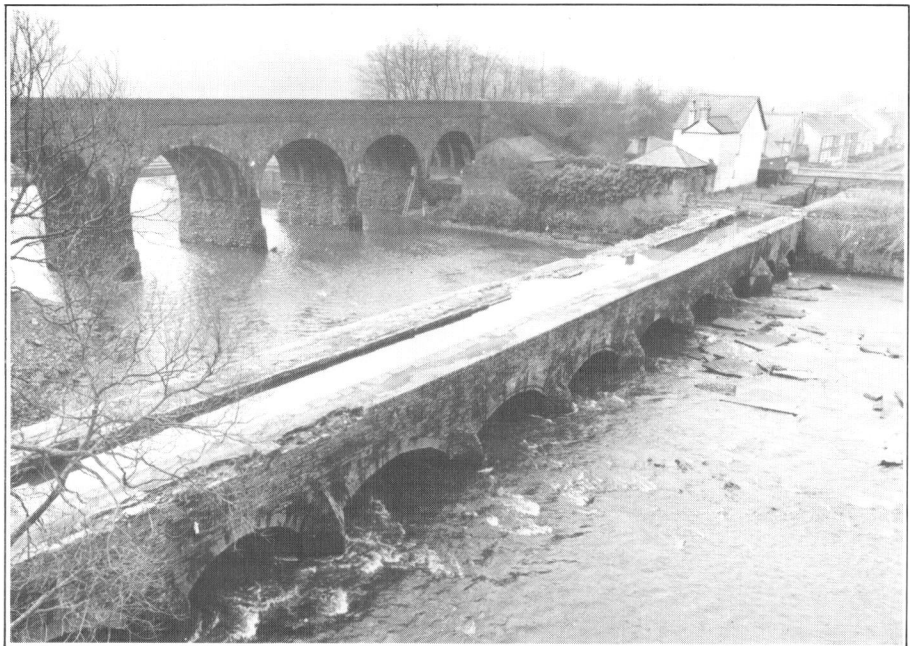

95 Aberdulais Aqueduct, Tennant Canal.

base remain. On the west bank of the river the coal was trans-shipped to barges on the branch canal. The loading dock and dry dock survive.

For further sites associated with the Swansea canal, see entries 18, 27, 38, 43 and 109.

84 SMITH'S CANAL
★   N I PS   □

Built by the coalowner John Smith in 1783-85 to replace the wagonway of Chauncey Townsend as the principal means of transporting coal raised in his Llansamlet colliery to the river. It consisted of a single pound, three miles long, and remained in use until the 1850s. Two sections where it has survived in a reasonable condition are the section above Llansamlet church [SS 6860 9768 to SS 6935 9788] and beside the smelting works at Upper Bank, Middle Bank and White Rock [29, 30], a section which includes an interesting cut-and-cover tunnel. The southern terminus was at Foxhole [SS 6630 9409] where the remains of stone-built coaling stages can be seen.

PENCLAWDD CANAL

A four-mile long canal built 1811-14 to carry coal for shipping at Penclawdd. How long it remained in use is uncertain. Much of its course was built over by the Penclawdd branch of the Llanelly Railway in the 1860s.

85 GOWERTON LOCK
☆   N I S   [SS 5934 9662]   □

One of four locks on the canal. It was converted into an air-raid shelter during World War II. The entrance is to be found under a lock-tail bridge now standing in waste ground. The dimensions of the lock suggest that Penclawdd Canal boats were about the size of conventional narrow-boats.

86 CANAL DOCK AND SCOURING BASIN, PENCLAWDD
★   N I S   [SS 5526 9609]   SC

The silted remains of the terminal basin are clearly visible on the former edge of the sea, the line of which has altered considerably since the 19th century. An arm of the canal ran out alongside the dockside quay and a culvert can still be seen that fed water from the canal into the scouring basin. Tidal river docks with enclosed basins or reservoirs on their landward side to scour the dock periodically by the opening of a connecting sluice were a common feature of the early 19th century docks in south Wales. Other examples existed at Port Tennant [SS 6723 9289], the Birmingham copperworks dock [SS 6691 9664], both in Swansea, and at Burry Port [75].

NEATH CANAL

Built 1791-96, largely with money raised by the townspeople of Neath, in order to exploit the undeveloped coal reserves of the upper Neath valley. Traffic ceased in the 1920s and it is now used as a source of water for local industries.

87 ABERPERGWM AQUEDUCT
★   N I S   [SN 8636 0585]   **LS**

A fine cast-iron aqueduct carrying the Nant Pergwm over the canal. Cast at Neath Abbey ironworks [45] in 1835 to the same pattern as the Resolven aqueduct [90]. Just to the south-west are the substantial remains of lock no. 15, Granary Lock [SN 8619 0570], and a lock-tail bridge.

88 MAES GWYN BOATHOUSE, MILL AND LIMEKILN
★   N I PS   [SN 8579 0516]   **LS**

Around the remains of lock 13 are a variety of interesting remains. An early 19th century lime-kiln is built into the lock platform [SN 8588 0517] and just downstream from the lock, at the Maes Gwyn overbridge [SN 8581 0512] are the remains of a pleasure boathouse built for the owners of the now demolished Maes Gwyn House. The foundations of a mill and a millstone also survive in the outbuildings of the mansion [SN 8579 0510]. It was driven by by-pass water from the lock.

89 CWM GWRACH BRANCH CANAL, WHARF-HOUSE AND LIMEKILN
★   N I PS   [SN 8549 0460]   ☐

A branch canal was built in 1817 across the valley floor to join a railway from Cwm Gwrach collieries. At the junction [SN 8553 0456] is a derelict wharf-house and at the terminus of the branch canal are the remains of a lime-kiln [SN 8700 0497].

90 RESOLVEN LOCK AND AQUEDUCT
★★   N I PS [SN 8272 0313]   Aqueduct: **LS**

Several features remain around lock no 8, Resolven Lock. They include the site of a riverside dock; the stump of a timber trestle tramroad bridge built over the River Neath in 1840-43; and a lengthsman's house [SN 8300 0326]. To the south of the road into Resolven is a cast-iron aqueduct, identical in design to that at Aberpergwm [87], which carries a stream over the canal. The next lock down, no 7, Farmer's Lock [SN 8245 0300] has a tail-lock bridge.

91 RHEOLA BROOK AQUEDUCT
★★   N I S   [SN 8422 0397]   **LS**

The most elaborate of the three cast-iron aqueducts that carried streams over the canal. Like those at Aberpergwm [87] and Resolven [90], it was probably cast at Neath Abbey ironworks [45] in the 1830s. The two side panels have eleven Tuscan columns cast into them and the deck of the trough is cast in eight plates fixed together by bolts through their surrounding flanges.

92 YNYSBWLLOG AQUEDUCT
★   N I S   [SN 8033 0110]   ☐

This was a fine and substantially intact six-arched aqueduct until a flood in 1980 carried away the larger part of two arches and left the present ruin.

93 NEATH CANAL DEPOT, TONNA
★★   N PS   [SS 7697 9879]   **CA**

The most complete of any of the depots of the south Wales valley canals. It is located on either side of the canal at lock no 1. The buildings consist of the canal manager's house and stables on the eastern side of the canal. On the western side is a long timber storage shed with a sawpit inside, also an open shed with block and tackle hoisting gear where lock gates were made. Attached to this is a store and a smithy.

For sites associated with the Neath Canal, see also entries 22 and 167.

TENNANT CANAL

Built 1820-24 by George Tennant to open up the estate around Neath which he had purchased in 1816. The engineer was William Kirkhouse. The canal started at a junction with the Neath Canal at Aberdulais and ended at Port Tennant on the eastern side of the River Tawe at Swansea. An earlier scheme of Tennant's, superseded by the canal, was the sea-lock at Red Jacket Pill. Traffic was carried until the 1930s. The canal is now used only to supply water to local industries.

94 RED JACKET PILL LOCK
★★   N I S   [SS 7259 9466]   ☐

A barge lock built by William Kirkhouse in 1817-18 to allow vessels of up to 50 or 60 tons to lock out of the Neath River and reach Swansea by means of an earlier canal constructed in the 1780s. The stub of this canal can be seen passing under a bridge beside the barge lock. Nearby is an interesting canal junction bridge and overflow from the later line of the Tennant Canal.

95 ABERDULAIS AQUEDUCT AND BASIN
★★   N I S   [SS 7725 9926]   ☐

Constructed in 1823 to complete the Tennant Canal. The 340' (104m) aqueduct carried on ten masonry arches is continued to the north-west by a cast-iron trough over an earlier navigable cut. The basin has remains of sunken boats and at the north a buried dry dock by a slip-way. At the junction of the Tennant and Neath Canals is a roving bridge, Pont Gam ('crooked bridge'), its flanking walls carried on a corbelled series of masonry courses. The only lock on the canal's main line is south of the aqueduct with office and lock-keeper's house.

96   CLYDACH AQUEDUCT, NEATH ABBEY

★   N I S   [SS 7388 9729]   ☐

A twin-arched masonry aqueduct with its arches submerged in the River Clydach and hence acting as an inverted syphon, built in 1821. The sluices with rusticated stone pillars to the south-east were built as part of the ill-fated Neath floating dock scheme [69]. The overbridge to the west of the aqueduct has parapets formed of cast blocks of copper slag with Roman numerals cast in.

97   SKEWEN RIVER DOCK AND LOCK

★★   N I S   [SS 7325 9677]   ☐

One of three river locks on the Tennant Canal. The fine rusticated masonry indicates that it was built as part of the Neath harbour scheme [69]. There is a large rectangular dock basin at canal level and the remains of timber loading stages project into the river.

98   SKEWEN CUTTING

★★   N I PS   [SS 7314 9691 to 7334 9725]   **SC**

A deep cutting created in 1821. During construction enormous problems were encountered with quicksand and part of the canal bed is laid in an inverted arch with vertical walls to seal the troublesome layer off on either side. A high stone causeway carried the Mines Royal Copperworks tramroad over the centre of the cutting.

99   KYMER'S CANAL

★   N

The first canal of any length in Wales, it was built in 1766-68 to carry coal from the Gwendraeth Fawr to Kidwelly where a quay was constructed [77]. It was later incorporated into the Kidwelly & Llanelly Canal [100–04] and subsequently a branch of the Burry Port & Gwendraeth Valley Railway was laid on the towpath in 1873. A section of the canal can be seen at Muddlescombe [SN 4270 0594 to 4097 0602], and at Pwll y Llygod a fine tramroad bridge crosses the canal [125].

KIDWELLY & LLANELLY CANAL

Incorporated in 1812 and built in fits and starts between 1814 and 1837. There were two distinct sections: the first followed the Gwendraeth Fawr from Kidwelly up to Cwmmawr and incorporated Kymer's Canal [99]. The second ran south-eastwards to Burry Port [75]. Initially traffic was shipped through Kidwelly, but later Pembrey and eventually Burry Port became the most important outlets. One of the most interesting features of the Gwendraeth Fawr section is the use of the three inclined planes rather than locks to raise the level of the canal: these were built in 1833-37 on the recommendation of the engineer James Green. In 1865 the company

94 Red Jacket Pill Lock, Tennant Canal.

35

converted itself into a railway company and the following year amalgamated with the Burry Port Company to form the Burry Port & Gwendraeth Valley Railway [151]. The line built by this company was laid to a great extent either on or beside the towpath of the canal.

100　SPUDDER'S BRIDGE AQUEDUCT
☆☆　　N I S　　　[SN 4276 0530]　　□

A massive piece of stonework with culverts that are barely adequate for the flow of water in the Gwendraeth. A plaque dates it to 1815 and until recently it was in use as a railway bridge.

101　　　　　　　　　PEMBREY BRIDGE
★　　N I S　　　[SN 4336 0098]　　□

102　　　　　　　　　STANLEY'S BRIDGE
★　　N I S　　　[SN 4278 0112]　　□

Rope marks are visible on the sides of both bridges, also the towpath, hacked away to provide clearance for the railway. Since the closure of this section of railway in 1985 it has reverted very much to its original status as a canal.

103　HIRWAUNISSA INCLINED PLANE
☆　　N PS　　　[SN 5229 1207]　　□

This is the only one of the three inclined planes that can still be seen, since at this point the railway followed an alignment alongside rather than on the incline in order to ease its gradient. The stone-built incline head stands in the undergrowth. Of the two other inclines, Ponthenri [SN 4764 0923] has been overbuilt by the railway—but note the name of the public house by the bridge—and Capel Ifan [SN 4915 1045] has been lost in later colliery buildings.

104　★　　　　　　　　　　　　CWMMAWR

The Canal terminated at Cwmmawr in a walled basin [SN 5311 1264] with an aqueduct across the Gwendraeth Fawr [SN 5311 1264] to a reservoir at Cwm y Glo.

102 Stanley Bridge, ex-Kidwelly & Llanelly Canal.

PRE-LOCOMOTIVE TRAMWAYS AND RAILWAYS

The first reference to a railed transport system in the Swansea area is from Neath where, in 1697, Sir Humphrey Mackworth started to lay down several short lines to bring the coal directly from the coalface to his smelting houses and wharves. The line remained in use in one form or another until the early 19th century and its ghost can still be seen in the street pattern in the Melincryddan district.

In Swansea coal was also being worked intensively by the same period, although there is no positive evidence for any wagonways having been constructed until Chauncey Townsend built one from Llansamlet to Swansea in the 1750s of which only slight traces remain. Coal was instead conveyed in carts or by packhorse along specially maintained coal roads, such as Councillor Price's Coal Road on the western side of the valley or Bussy Mansel's Great Coal Road on the east. But with the growth of the coal industry from the mid-18th century in response to the demands of smelteries, railed transport systems soon became widespread.

The early wagonways were of course constructed with wooden rails, perhaps reinforced with iron strips on the running surfaces, but in 1776 John Morris introduced iron rails to Swansea: 'In Nov'r 1776', he records, 'I wrote to Messrs. Darby & Co at Coal Brook Dale that I had sent them a pattern of wood about 4 ft long and 5 in. wide, to cast Iron plates for wheeling Coal in my Collieries, each plate to weigh ab't 56 lbs'. The experiment was a success and by 1788 there were '240 tons of Cast Iron Tram Plates underground in Landore Coll'y', or enough for about 3¾ miles of railway. Just conceivably, Morris may have anticipated John Curr of Sheffield in inventing **L**-section tramplates, in 1787.

However, by 1800 the **L**-shaped tramplate had become widely accepted and was more or less standard practice in south Wales. For the next thirty years it is rare to find any edge-rail systems, although the occasional one did exist, such as a Losh-rail line at Clydach. It was not until the 1830s that tramplates yielded to edge-rails once again and new lines were built as railways rather than tramroads.

Until the advent of the modern locomotive railway the canal was regarded as the main trunk artery of communication and tramroads were almost always seen as mere extensions of or substitutes for a canal. They were built in situations either where the amount of traffic on offer was not enough to justify a canal, or where the terrain was such that canal construction and operation would be prohibitively difficult and expensive. There were lines that were independent of canals, but they were few. The longest of them was the Carmarthenshire Railway, but even that was subsidiary to water transport, since its main function was to carry coal to the dock in Llanelli. Even such lengthy lines as the Brecon Forest Tramroad remained firmly dependent on their canal, and as late as 1840 new lines were still seen as tributaries to canals, such as Parsons', Mynydd Newydd, Pentrefelen or Gwaun-cae-Gurwen.

Tramroads were almost always operated by horses on the level, but the nature of the country is such that the use of inclines was common, and very often these inclines, either powered or self-acting, form the best surviving features. Locomotives are known on Scott's Tramroad in 1819 and again in 1833 and at Cwmavon in the 1840s. There is just a faint possibility that a Blenkinsop rack locomotive might have been used at Landore in Swansea in about 1814. Claypon's extension of the Brecon Forest Tramroad was certainly engineered with the use of locomotives in mind although it is questionable whether they were ever used.

Inevitably most of the once extensive mileage of tramroads in the Swansea region has been obliterated by subsequent development, but there are still remains which repay investigation, generally because of their remote location. Of these the Brecon Forest and Parsons' must be the best. In other cases the modern street pattern has been created along the lines of the former tramroads. The example of Neath has already been mentioned, but another striking case is in the Landore/Brynhyfryd area of Swansea.

105   SCOTT'S TRAMROAD
★   N I   ☐

Built 1817-19 to carry coal from Scott's Pit [4] to his shipping place on the River Tawe at White Rock. A locomotive—quite probably built by George Stephenson—is known to have worked the line for a short period., c1819-20. The tramroad passed into the possession of C H Smith along with the colliery in 1828 and he is also known to have been using a locomotive in 1833. The tramroad was acquired by the Swansea Vale Railway [128] in 1845 and formed the basis of their line. The chutes through which coal was discharged can be seen at White Rock [29], just downstream from the recently re-excavated dock [SS 6630 9470].

106
MYNYDD NEWYDD RAILROAD INCLINE
★   N I

107   PENTREFELEN RAILROAD
★   N I

The Swansea Coal Company was formed in 1839 by two major copper- smelting houses to take over the Penyvilia Vein Co and so ensure their supplies of coal. The business they acquired was run down and the existing pits close to exhaustion. One of their first tasks therefore was to sink two big new pits, Mynydd Newydd [SS 6388 9648] and Pentrefelen [SS 6573 9910]. Both were connected by railroad to the Swansea Canal and so to the Hafod and Morfa works [27, 28]. Pentrefelen started to yield in 1844 and Mynydd Newydd in 1843. The railroads remained in use until the collieries closed in 1893 and 1932 respectively.

The only remaining feature of the Mynydd Newydd railroad is part of the incline that lowered coal into the Swansea Valley ★ [SS 6525 9590]. The Pentrefelen railroad can be traced for most of its route in the form of rough tracks or linear features. It terminated at a wharf beside the Swansea Canal at SS 6721 9847. One well-preserved section is where it crosses Morriston golf course on a slight embankment [SS 6525 9891]. Two-holed sleeper blocks can be seen at various points along both lines.

108   GWAUN-CAE-GURWEN RAILWAY
☆   N I PS   ☐

Roger Hopkins leased some 700 acres of coal at Gwaun-cae-Gurwen in 1837 and the following year made a start on the construction of a railway from his colliery [SN 7189 1191] to the Swansea Canal at Pontardawe. Only the first mile of this six-mile line was completed before Hopkins realised that his interests would be better served by the Llanelly Railway [146-49], then in the process of building a line up the Amman valley. Work on the canal-linked railway was abandoned and Hopkins extended the LIR's line to his colliery in 1840-41. The northernmost parts of the abandoned works of 1838-40 have been lost to land reclamation but an impressive embankment and cuttings remain behind the present village [SN 7070 1114 to 7105 1153].

109   YARD BRIDGE
★   N I S   [SN 8154 1248]   SC

A large stone arch with pierced spandrels that carried a tramroad over the gorge of the River Tawe. It was built c1824 by Daniel Harper to give access to his Lefel Fawr on the south-east bank of the river. Lefel Fawr was opened c1796 and was originally intended to be an underground branch of the Swansea Canal. Its rock-hewn portal can still be seen. The early 19th century house beside the bridge was the colliery manager's office and home. Just downstream, on the south-east bank can be seen the stone abutment of an earlier timber trestle bridge.

BRECON FOREST TRAMROAD

Constructed 1821-25 by the London indigo merchant, John Christie, to open up a 15,000 acre holding which he had purchased from the Crown in 1819. It linked collieries at Onllwyn and Gwaunclawdd with limestone quarries at Penwyllt and depots at Cnewr and Sennybridge. Following the bankruptcy of Christie in 1827 the tramroad passed to his chief creditor, Joseph Claypon, a Lincolnshire banker, who built an extension to the Swansea Canal in 1832-34. The BFT remained in existence until the 1860s when it was replaced by the Neath & Brecon Railway [135].

110   SENNYBRIDGE DEPOT
☆☆   N I S   [SN 9170 2833]   ☐

The terminus of the BFT from 1823 until its closure in 1867. The most obvious features are two limekilns and a row of houses. The tallest of these was a warehouse constructed in 1831; the others housed stables, the tramroad manager and a weighbridge. Various earthworks in the field mark the course of the tramroad and ramps for a cart road, and the dam and pond of the short-lived Devynnock Foundry.

111   CLYDACH CAUSEWAY
☆☆   N I S   [SN 9158 2710]   ☐

A large stone-walled causeway which carried the tramroad over a ravine. It is now so ruinous

109 Yard Bridge.

that it clearly reveals the details of its construction and is badly in need of immediate consolidation. It also carried a carriage road as part of Christie's attempt to improve every form of communication over the mountains.

112 CNEWR DEPOT
☆☆   N I S   [SN 8903 2206]   □

A model farm constructed 1821 as the centre for sheep farming in the newly enclosed Great Forest. Also the original depot of the BFT. Its quadrangular plan is similar to that of the other two contemporary model farms of Belfont [166] and Forest Lodge [165]. The keystone of the entrance arch has a carved tramroad wheel and the initials 'J C' for its builder, John Christie. The houses on either side were probably for the agricultural and tramroad managers. Inside, the other three sides of the quadrangle were occupied by a lime store, stables and a wool store.

113 GRAWEN DEPOT
☆☆   N I S   [SN 8865 2082]   □

Developed c1830 as the central depot on the BFT in place of Cnewr. The ruins of stone-built cottages and stables remain on either side of both the tramroad and a partially completed carriage road which ran alongside it. Extra building terraces were created but never used: instead, in the 1830s, cottages were built on the line of the carriage road. The foundations of these survive, and to their east, the ruins of a covered lime shed. On the hillside above the depot is the course of the first formation of the tramroad constructed 1820-21 but abandoned almost as soon as it was built in favour of the present route.

114 BWLCH BRYN-RHUDD CAUSEWAY
☆   N I S   [SN 8667 1868]   □

An almost complete tramroad causeway survives near the mountain pass at Bwlch Brynrhudd and is incorporated into the formation of the Neath & Brecon Railway [135]. It is now accessible for most of the year (apart from the lambing season) as part of a footpath to Penwyllt quarries [115].

115   **PENWYLLT**
★★   **N I PS**   [SN 8565 1573]   ☐

This large complex of remains which surround the still-functioning limestone quarries includes:-
— tramroad ramps and limekilns [SN 8556 1613].
— Victorian back-to-back houses converted from tramroad workers' cottages [SN 8535 1588].
— Craig-y-nos railway station [135].

These are all to the north of the present quarries. To the south are a fine Victorian workers terrace, railway formations and the kilns of the Penwyllt silica sand brickworks [SN 8546 1518], and nearby a nearly complete block of railway-age limekilns. On the mountainside to the east the nature reserve is intersected by an unfinished tramroad incline, probably part of an abortive scheme of Christie's [SN 8561 1540 to 8614 1571], and by a narrow-gauge tramway which served the brickworks. A standard gauge zig-zag formation can also be seen descending from the summit of the mountain [SN 8582 1635 to 8630 1619].

116   **CLAYPON'S INCLINE AND ENGINE-HOUSE**
★★   **N I S**   [SN 8013 0942]   **SC**

Built by Joseph Claypon, 1832-34, as part of his extension to Christie's original line. It was primarily intended to supply limestone from Penwyllt to Ynysgedwyn ironworks [42], but Claypon also intended it to play a part in opening up the newly acquired Great Forest which he was trying to develop. The formation of the incline [SN 7880 0795 to 8006 0944], the basement of the engine-house and the succession of cuttings and embankments over the Drum Mountain are a public footpath and as scheduled monuments are safe from the encroachments of opencast working, at least for the time being.

112 Cnewr Tramroad Depot and Farm.

115 Letterhead of Penwyllt Dinas Silica Brick Co.

PARSONS' RAILWAY

Parsons' railway—otherwise known as Parsons' Folly or the Glyncorrwg Mineral Railway—was built in 1839-42 to open up a vast area of hitherto inaccessible coal in the area behind Neath. It ran for 7½ miles from the Neath Canal at Aberdulais [95] to the small Blaencregan level [SS 8555 9895]. The original owners were obliged to surrender possession in 1843 but regained it a few years later when their successors ran into difficulties. The whole project was unrealistic and the railway was lifted in 1852. It was a spectacular piece of engineering and much of the route can be followed without difficulty.

117 INCLINES 1–4

From the canal, the railroad climbed the side of the Neath valley by a series of four self-acting inclined planes. The stone-built causeway which carries the final stages of incline 4 is a noteworthy feature ★★ I [SS 7870 9855] ☐. Lower down a large arched bridge still exists ★★ [SS 7801 9885] ☐.

118 SUMMIT LEVEL
★★   I   [SS 7876 9853 to 7919 9781]   ☐

The distance is about half a mile. It passes through a deep rock cutting and over a massive embankment. At the north-west end it intersects two early reservoirs built for Mackworth's Melincryddan works in Neath.

119 INCLINE 5

The remains of the engine house for what was the only powered incline on the railway can still be seen ★★ I [SS 7917 9782] ☐. A short distance to the west lies the dam of the reservoir that provided water for the engine ★ I [SS 7915 9770].

120 TONMAWR VILLAGE

The modern village is built along the line of the railway and post-dates it. Features to be noted include Parsons' first level ★ [SS 8008 9693] ☐; the remains of a railway-charged limekiln ★ [SS 8033 9627] ☐; and bridge abutments over the road ★ [SS 8038 9632] ☐.

121 FFORCHDWM VIADUCT
★   N I S   [SS 8157 9692]   ☐

The piers of a fine four-span stone viaduct built in 1842. The arches were demolished in 1979. The road from Tonmawr to this point is built on the railway formation.

122 GLYNNEATH INCLINED PLANE
★★   [SN 8913 0654 to 8992 0631]   ☐

A half-mile long incline on the Cefn Rhigos tramroad, built 1803-05 to connect the ironworks at Aberdare with the Neath Canal. It is the second oldest steam-operated incline known to have been built in the UK—and presumably the world. It was worked—although not very successfully—by a high-pressure engine. The site of the engine house at the head of the incline is clearly visible.

121 Fforchdwm viaduct.   *Photograph: P R Reynolds.*

123 BRYN RAILROAD

Constructed 1839-40 to carry coal from the area around Bryn [SS 8150 9230] to the iron and copper works of the English Copper Co at Cwmavon. It was first used in 1841 and locomotive power was introduced on the level section in 1845. Much of the route can still be traced. The River Afan is crossed by a three-arch bridge at Ynysafan ★★ [SS 7870 9231] □. The incline up the side of the Afan Valley can be identified and derelict masonry at the summit marks the site of the engine house ★ [SS 7911 9192] □. From this point to Bryn the modern B4282 road is either built on or directly alongside the railroad formation. Stone blocks can be seen in places.

124 STANLEY'S TRAMROAD, PEMBREY
★   N I   □

Built c1820 to carry coal from a pit at Pembrey [SN 4330 0128] to Pembrey harbour [76]. It probably remained in use until c1863 when the pit was flooded. For much of its length of one mile it is carried on an embankment which forms a prominent feature where it runs across Ashburnham golf course ★ [SN 4345 0165]. The embankment was subsequently pierced by the Kidwelly & Llanelly Canal [102].

125   PWLL Y LLYGOD BRIDGE
☆☆   [SN 4461 0681]   □

A fine single-arched bridge of c1770 which carried a tramroad from coal pits near Pwll y Llygod farm over the Gwendraeth Fawr to Kymer's Canal [99]. The collieries—and presumably the bridge—were still in use in the 1860s. It is a monument of some considerable importance, being the oldest railway bridge in Wales and the second or third oldest in the UK—and hence the world.

For further sites with remains of tramroads, see entries 2, 5, 20, 31, 83, 90, 98, 152 and 168.

127 Landore viaduct as originally constructed, from the south.   *National Library of Wales*

PUBLIC LOCOMOTIVE RAILWAYS

Although south Wales had a high mileage of horse-worked tramroads, locomotive railways made a late appearance in the region. The Llanelly Railway (1839) is generally regarded as the first modern railway in Wales, although in fact it had as much in common with earlier 'hybrid' railways at first as it had with later modern railways.

The first trunk line was the South Wales Railway, opened from Chepstow to Swansea in 1850, although it was not until 1852 that a physical connection was made with the national network. With the opening of this base line feeders soon started to be built up the main valleys. At first these routed traffic onto the SWR but soon sought their own access to the docks at Swansea and elsewhere.

The SWR and most of its feeders were built to the broad gauge and this hampered the development of rail-borne traffic between south Wales and its main markets in the Midlands and north of England. The 1860s therefore saw attempts by English standard-gauge companies to acquire access to south Wales. The process was welcomed by industrialists in south Wales and eventually three standard-gauge outlets were completed from Swansea—the Vale of Neath line to Pontypool Road (1864), the Central Wales line to Craven Arms (1868) and the Swansea Vale line to Hereford in 1873. With the conversion of the South Wales main line to standard-gauge in 1872 and the absorption of small local companies by larger English companies, services in the Swansea region settled down into a pattern which was to last until 1923—the Great Western Railway provided the principal access but other important extra-regional connections were also supplied by the London & North Western and Midland companies.

A number of small local companies also existed which were heavily dependent on coal traffic. Thus the anthracite coalfield in Carmarthenshire was served by the Llanelly & Mynydd Mawr (opened 1883) and the Burry Port & Gwendraeth Valley Railways (opened between 1869 and 1886), while in the east the Rhondda & Swansea Bay and Port Talbot Railways were both promoted to attract traffic to new docks at Swansea and Port Talbot. The Neath & Brecon also retained its independence, although deeply involved with the Midland Railway.

The beginning of the 20th century was the period at which the south Wales coalfield was at its most productive. To the north and west of Swansea anthracite showed a marked increase, and in response to this the GWR promoted several ambitious new railway schemes to improve the flow of traffic to Swansea docks. These plans were only partially realised before the outbreak of World War I put an end to further construction.

The railway network in the western part of the coalfield contracted slowly during the inter-war years, but no substantial sections were closed until the 1960s. Today the South Wales main line carries a reasonably heavy passenger and goods traffic. The Central Wales line (Llanelli—Shrewsbury) remains open for passenger traffic, and the demands of the coal industry have led to the retention of a number of freight-only branches.

SOUTH WALES RAILWAY

Incorporated in 1845 and built to a survey by Brunel. It followed an undemanding route across the coastal plain which called for few great works of civil engineering. The first section between Chepstow and Swansea was opened in 1850 but it was not until 1852, with the completion of the Wye bridge at Chepstow that south Wales had an unbroken link with the rest of the network. It was completed as far as Neyland in stages by 1856 and taken over by the Great Western Railway in 1863. Converted from broad to standard gauge in 1872.

126 LLANSAMLET FLYING ARCHES
☆   **N**   [SS 7018 9747]   □

Shortly after the line was opened, in 1851, the sides of the cutting started to slip. As a countermeasure, to resist further movement, Brunel had four flying arches built across the cutting, weighted with copper slag.

129 Loughor viaduct after restoration. Photograph: S K Jones.

127 LANDORE VIADUCT
★★   N I PS   [SS 6622 9580] ☐

The original viaduct across the Tawe valley, built in 1847-50 was one of Brunel's timber structures with some masonry piers. It was replaced in 1888-89 by a steel girder bridge and the eastern part converted into an embankment. The decking was renewed in 1978-79 but without substantially altering the appearance of the viaduct. The five masonry piers, each with two arched openings, which support the bridge where it crosses the river, belong to the original structure.

128 SWANSEA (HIGH STREET) STATION
★   I   [SS 6574 9360] ☐

The present facade was built by the GWR in 1934-35 and is typical of that company's work of the period. Behind the terminal block, on the western side of the station, part of the second station of 1877-78 survives.

129 LOUGHOR VIADUCT
★   I S   [SS 5605 9801] ☐

The only surviving timber viaduct to have been designed by Brunel. As built it consisted of 17 timber pile-driven piers with a superstructure of timber trusses and one wrought iron opening span. This latter was replaced by a fixed span in the late 19th century and the timber superstructure was replaced with iron, and later steel, girders. In 1981 new timber piles were driven in and jointed to the originals which has had the effect of changing the configuration of the piers. The original timber was yellow pine from Memel on the Baltic Sea.

130 SWANSEA DISTRICT LINE AND ASSOCIATED BRANCHES

Construction started in 1907 and the line was opened in 1912-15 from Court Sart Junction (Briton Ferry) to Morlais Junction (Llangennech). The most eastern section used the Rhondda & Swansea Bay Railways's viaduct to cross the Neath river [138]. The line forms a railway by-pass to Swansea and Neath and was promoted by the GWR for two purposes: to improve access to Swansea docks from the expanding anthracite coalfield and to provide a fast uncongested route for expresses to Fishguard. Because of this, gradients were kept as easy as possible and the curves laid to a very generous radius. Consequently most of the line is carried in cuttings or tunnels or on bridges and embankments. In themselves all these features are impressive but none are of particular importance. Considered as a whole the line makes a powerful statement about the ideas which underlay railway promotion and construction in the early 20th century. The Fishguard traffic never developed on the scale anticipated, but the line still serves a useful purpose as a freight carrier.

In 1913 a branch was started from Felin Fran [SS 6917 9857] to Gwaun-cae-Gurwen to tap the anthracite traffic of the Amman valley. Work started at both ends but was suspended in 1915. After the war lack of funds prevented completion of the line. The southern stub runs as far as Clydach where it serves the Mond nickel works [35]. The formation was completed for a further two miles but the rails were never laid. At Pontardawe, in the Uplands district, a short tunnel remains [SN 7180 0407] and this marks the furthest point to which the earthworks from the south extended. For the northern stub, see entry 149.

131 SWANSEA & MUMBLES RAILWAY

The Oystermouth Railway, as it was originally known, was incorporated in 1804 to carry coal from the Clyne Valley and limestone from Mumbles. In 1807 a passenger car started to operate, claimed to be the first regular passenger service by rail. The line was relaid as an edge railway in 1855 and steam trains operated from 1876. It was extended to Mumbles Pier in 1898. Electric trams replaced steam trains in 1929 and the line closed in 1959-60. The only remaining structure is Blackpill Electric Sub-Station ★ [SS 6191 9057] ☐, built in 1928 as part of the process of electrification. Mainly constructed of brick, with large round-headed windows and a stone colonnade (originally Blackpill station) which adds to its character. At present the building is disused.

132 MIDLAND RAILWAY (SWANSEA VALE SECTION)

The Swansea Vale Railway was formed in 1845 to buy up and extend Scott's tramroad [105]. They eventually built a line as far as Brynamman which was opened in stages in 1852-64. A loop through Morriston followed in 1871-75. Meanwhile the Midland Railway acquired the company and used its line as part of a route from Hereford to Swansea. Passenger traffic ceased in 1950 and the line closed in stages in 1964-83. Few remains of any significance survive. At Six Pit [SS 6827 9567] the modern Swansea Vale Railway, a preservation group, has its headquarters: the intention is to re-open the 1½-mile section between Six Pit and Upper Bank.

VALE OF NEATH RAILWAY

Incorporated in 1846 to link the iron and coal producing area of Aberdare and Merthyr Tydfil with the ports of Neath, Briton Ferry and Swansea. Engineered by Brunel, it called for ferocious gradients to lift the line over the watershed between the Neath and Cynon valleys. It was opened from Neath to Aberdare in 1851 and subsequently extended to Pontypool Road to provide a link with the English Midlands. An extension to Swansea was opened in 1863. Acquired by the GWR in 1865. Closed in stages between the 1960s and the 1980s.

133 ABERDULAIS VIADUCT
★ I [SS 7729 9325] ☐

A late 19th century masonry viaduct of five arches which replaced Brunel's original timber structure. Close observation reveals the difference between the late 19th century masonry and that of the abutments of Brunel's bridge.

134 SWANSEA (NEW CUT) BRIDGE
★ [SS 6615 9318] ☐

Four massive stone piers stand in the river bed. They carried a bridge which opened in 1863 and closed in 1965.

135 NEATH & BRECON RAILWAY

A typical contractor's railway, the contractor in this case being John Dickson. Opened from Neath to Onllwyn in 1864 and on to Brecon in 1866. Between Onllwyn and Sennybridge it closely followed the course of the Brecon Forest Tramroad [110-6]. In its early days Dickson operated traffic with the first two Fairlie locomotives ever built. A connection with the Swansea Vale Railway [132] was made in 1873 which allowed it to form part of the Swansea-Hereford through route. The line north of Onllwyn closed in 1962-81 but south of Onllwyn it remains open and carries traffic from collieries and washeries in the Dulais valley.
At Craig-y-Nos station ★ [SS 8546 1572] ☐, rather better facilities exist than might have been expected in this remote quarrying community [115], because this was the station used by the celebrated Victorian singer Adelina Patti, who lived nearby at Craig-y-Nos on the valley floor. It has been refurbished by Messrs Hobbs (Quarries). *See also entry 162 (Dickson's Row)*.

RHONDDA & SWANSEA BAY RAILWAY

Promoted by Swansea interests to attract coal traffic from the Rhondda Fawr to the newly opened Prince of Wales Dock. It was incorporated in 1882, but construction was not finally completed until 1900, the original plans having seen frequent modification. The Afan valley section was closed between 1962 and 1970, but the Neath River crossing and approaches to Swansea remain in use.

136 CROESERW VIADUCT
★ [SS 8643 9605] ☐

Opened 1890 and closed in 1959 when traffic was diverted on to the adjacent GWR line in order to save the cost of repairs.

137 CYMMER STATION
★ [SS 8610 9605] ☐

The refreshment room of the RSBR station (opened 1885) survives as a public house, 'The Refreshment Room' or, colloquially, 'The Refresh'.

138 NEATH RIVER BRIDGE 45
★★ N [SS 7314 9635] ☐

Constructed 1892-94. The only oblique swing-bridge in the UK. The total length is 388'

(118m) of which the swinging part, which can be opened to give access to Neath for vessels on the river, is 180' (55m) long. There are five fixed spans, the whole resting on steel cylinders sunk 40' (12·2m) into the river bed and filled with cement.

139   DANYGRAIG LOCOMOTIVE SHED
☆   [SS 6937 9318]   ☐

Opened in 1896 as the main locomotive depot and repair shops of the RSBR; closed as such in 1964. It now houses a chemical company. Built of local stone with yellow brick arches and corners. The most northerly five bays formed the locomotive shed; the central three bays the locomotive repair shop; and the four bays nearest the road, the carriage and wagon repair shop.

SOUTH WALES MINERAL RAILWAY

Incorporated 1853 to connect collieries at Glyncorrwg with the dock planned for Briton Ferry [68]. Engineered by Brunel and built, most unsuitably, on the broad gauge. The principal feature was a mile-long incline at Briton Ferry, even in the 1850s an anachronistic method of railway engineering. The line was opened in 1861- 63 and throughout its existence remained almost exclusively a coal-carrying line. The incline was closed in 1910 and the rest of the line between 1947 and 1970.

140   YNYS-Y-MAERDY INCLINE
★   [SS 7442 9484 to 7591 9521]   ☐

The incline can be followed in its entirety except at the downhill end. At the summit the overgrown site of the winding engine can be explored and evidence for the way in which the traffic was handled be deduced.

141   GYFYLCHI TUNNEL
★   [SS 8096 9600 to 8167 9527]   ☐

1109 yds (1015m) long. The western end was covered by a landslip in 1947 which led to the closure of the tunnel, but the eastern portal still stands high on the northern flank of the Afan valley amid surrounding forestry.

142   CYMMER VIADUCT
★   [SS 8575 9610]   ☐

Built by the GWR to link the SWMR to their Bridgend-Abergwynfi line. A seven-span viaduct with wrought-iron lattice girders, cross girders, timber decking and masonry piers. Opened 1878 and closed to traffic 1970.

46 PORT TALBOT RAILWAY & DOCK CO.

Incorporated 1894 to acquire and improve the earlier dock of 1837. Between 1895 and 1898 the new dock [70] was started and railways were then opened in three directions to draw traffic from the western central valleys.

143   CWM CERWYN TUNNEL
★   [SS 8321 9157 to 8381 9090] ☐

A single bore tunnel, 1012 yds (925m) long and completed in 1897 as the keystone of the west portal shows. It carried the line from Port Talbot to Maesteg though the Afan/Llynfi watershed. Closed 1964.

144   PONTRHYDYFEN VIADUCT
★   N   [SS 7930 9415]   ☐

A fine ten-arch viaduct which carried the South Wales Mineral Railway Junction line of the PTRD. Constructed 1897-98; closed 1964.

145   TONMAWR JUNCTIONS
★   [SS 7971 9637]   ☐

Earthworks of a complicated burrowing junction by which the PTRD made a connection with the South Wales Mineral Railway. The abstraction of traffic at this point made it possible to close the Ynys-y-Maerdy incline [140] in 1910.

LLANELLY RAILWAY & DOCK CO

Formed in 1828 to built a railway from St David's Colliery [13] to a new dock to be built at Llanelli [73]. Extended to Pantyffynnon with branches to Cross Hands, Brynamman and Gwaun-cae-Gurwen in 1842 in order to open up the anthracite coalfield. Further extended to Llandeilo (1857), Carmarthen (1865) and Swansea, with a branch to Penclawdd (1867). This last extension overstretched the company and in 1871 it was dismembered. The original 1833-42 lines were acquired by the GWR while the LNWR took over the Carmarthen and Swansea lines.

146   PONTARDULAIS TUNNEL
☆   I   [SN 5869 0390]   ☐

The oldest surviving railway tunnel in Wales, opened in 1839 and 88 yds (80·5m) long.

147   PANTYFFYNNON STATION
★   I   [SN 6227 1076]   ☐

A single-storeyed building built of stone in an Italianate style reminiscent of Brunel. Probably dates from 1853-57 when the LIR was extending northwards to Llandeilo. One of the more attractive stations in south Wales.

148   LLANMORLAIS STATION
☆   [SS 5335 9466]   ☐

A typical example of a small LNWR station, now used as a private house. Built in 1883 when the Penclawdd branch was extended further into Gower, as the iron numerals above the door show. Timber construction.

149 GWAUN-CAE-GURWEN VIADUCTS
★ [SN 7005 1209] ☐

Two large brick viaducts dating from the years immediately before World War I as part of a major GWR programme of construction in the anthracite coalfield to improve access to Swansea docks. The more northerly of the two now serves the opencast washery, but the southern one has never had track laid on it, since the outbreak of war brought construction to a standstill.

150 LLANELLY & MYNYDD MAWR RAILWAY

Incorporated 1875 to buy and reconstruct the Carmarthenshire Railway, a tramroad dating from 1802, which had been intended to open up the coalfield to the north of Llanelli. It never achieved the results that were expected of it and by 1830 most of it was derelict, although the southernmost section still continued to carry some coal to Llanelli docks. The LMMR reconstructed the line and opened it in 1883 as far as Cross Hands, 16 miles north of Llanelli. The new railway followed the original alignment closely, only departing from it in a few cases where the 1802 formation can now be seen as a separate feature. Throughout its existence it has depended almost exclusively on coal traffic and now serves the deep-level anthracite mine at Cynheidre. Its origins as a horse-worked tramroad are clear, in that it follows a winding course with the minimum of heavy earthworks.

151 BURRY PORT & GWENDRAETH VALLEY RAILWAY

Incorporated 1866 as a combination of the Burry Port Harbour Co and the Kidwelly & Llanelly Canal Co who, the previous year, had been authorised to convert their canal into a railway. The line was built on the canal bed or the towpath. It was opened to Pontyberem in 1869 and extended to Cwmmawr in 1886. In 1985 the route into Burry Port was closed and a new connection with the main line made at Kidwelly. It still carries regular coal traffic from the washeries at Cwmmawr and Carway. For surviving features, which are shared by the railway and the earlier canal, see entries 100–103.

ROADS

In the early 18th century (and possibly earlier) special coal roads were built by the local landowners and coalowners and their tenants to convey the produce of their mines to the wharves and docks along the navigable rivers. Part of the modern road network is based on these roads. In the 1820s substantial sections of road were built at a date contemporary with or shortly after that of some of the longer horse-railways. Thus the modern A4067 road round Swansea Bay follows the course of the realigned Oystermouth Railway and the trunk road between Swansea and Brecon (A4067 again) has the formation of its 1 mile summit section over the Brecon Beacons determined by the tramroad entrepreneur John Christie. Similarly the A4069 road over the Black Mountain from Brynamman to Llangadog was also largely built in this period by the Brynamman coalmaster John Jones to supply fuel to limekilns on the Black Mountain, although this was a road from the outset, not a tramroad.

Many of the public railways built under the terms of the Swansea Canal Act later became public roadways and much of the road network of the Swansea suburb of Landore is based on the formation of horse railways.

Turnpike roads were of course built in the area. A fine toll-house survives at Ynysderw near Pontardawe [SN 7168 0342] on the main road up the Swansea valley. There are also some interesting sections of abandoned road such as the six miles of mountain-top turnpike in the Brecon Beacons between Maen Llia [SN 9248 2445] and Forest Lodge [169], including a ruined toll-house and a bridge. A fine section of abandoned mountain-top road also lies alongside the A4067 at the mid-18th century Pont Gihirych bridge [SN 8858 2119], approached by a gradient of 1 in 7.

152 HAFOD BRIDGE
★　　N I S　　[SS 6578 9420]

An early 19th century bridge which carried the Swansea to Neath turnpike road over a small cwm. A road ran through the central arch, a tramroad through the eastern arch, and the Burlais brook, now culverted, through the western arch which is now barely visible. The tramroad was constructed c1790 to carry coal to Swansea harbour. It was rebuilt as a railway in 1848 and closed in the early 1870s. The bridge was widened c1878-79 when a street tramway was laid over it: the two phases of construction can be seen clearly from the underside.

153 PONTARDAWE BRIDGE
★　　[SN 7248 0371]

Now overshadowed by a modern concrete bridge, the original Pont ar Tawe was built by William Edwards (1719-1789), best known for his bridge at Pontypridd. The bridge, built c1770, has a single span of 80' (24m) and, like the bridge at Pontypridd, was originally built with cylindrical openings in each haunch to relieve the weight. These were filled in in 1819. Pontardawe is the only survivor of three bridges which Edwards built across the Tawe. The other two were lower down in the neighbourhood of Morriston: Wychtree bridge [SS 6737 9789], built in 1778 and demolished in 1959, was another single-span bridge. Beaufort bridge [SS 6709 9712] was built in the 1760s and demolished in 1968. For other work by Edwards in the Swansea valley see entries 67 and 157.

156 Morris Castle, Swansea.

HOUSING

A factor of great importance to the success of any industrial undertaking is a dependable supply of labour. At the beginning of the industrialisation of south Wales the region was thinly populated and in order to attract labour and concentrate it close to the works, one of the incentives an industrialist frequently offered was either ready housing or building plots on favourable terms. This was especially the case where skilled labour was concerned or in particularly isolated areas, but even in areas of established settlement an industrialist still found it worth his while to assist his workmen with housing. Two examples in Swansea are Morris Castle and Morriston, both developed by the coal and copper magnate, John Morris, in the late 18th century.

In more isolated places the provision of housing was almost essential and small rows or streets were run up as part of the development process. Nearly all of these have now been demolished, although surviving examples can be seen at Clydach, Onllwyn or Penwyllt. Also of interest is the long single-storey terrace at Seven Sisters, a type of housing which is typical of the early phases of development in a hitherto rural valley. Rather more upmarket are the unusual three-storey terraces in Cwmavon and Glyncorrwg which were built for managerial staff or skilled labour in the mid 19th century.

Generally speaking, housing on the western side of the coalfield was left far more to private enterprise than in the east. It was an older, more settled area and generally market forces could cope with the requirements for new housing which tended to be gradual rather than a sudden explosion of demand as was more often the case in the east. Typically in the west a man who came into a little money would build a few houses for letting by way of an investment. There are few of the long grim company terraces of the Mid Glamorgan valleys. The scale of development is more varied and the units are smaller. What company housing there is, is less dominating: examples include College Row, Ystradgynlais for Yniscedwyn ironworkers or Vivians Town, Swansea, for Hafod copperworkers.

154   VIVIANSTOWN (TREVIVIAN)
★   [SS 6583 9473]   **CA**

Two terraces of houses on the Neath Road were built for his workers by J H Vivian of the Hafod copperworks c1840. They are remarkable for their extensive use of moulded blocks of copper slag in the structure and especially as coping stones on the walls dividing the long front gardens. The terraces are substantially as built although modifications have been made to doors and windows. Further housing for the copperworkers extended rapidly behind these two terraces (note in particular Vivian Street) during the 1840s and in 1845-46 the Vivians built the Hafod Copper Works Infant School in Odo Street. St John's church, also on Odo Street was built by them, 1879-80. The street names all have associations with the Vivian family.

155   GRENFELL TOWN
★   [SS 6684 9498]   ☐

A few terraces of houses at Pentrechwyth, now much modernised, which were built for workers in the Grenfell family's Upper Bank and Middle Bank copperworks [30]. The street names have associations with the family. The Grenfells were a family with a strong social conscience and they established the Kilvey schools (c1806) and the adjacent All Saints church of 1842 [SS 6638 9410]. Both were surrounded by further housing which has since been demolished.

156   MORRIS CASTLE
★★   **N I**   [SS 6596 9640]   **SC**

An early block of industrial flats, built between 1768 and 1775 by the coal and copper magnate, John Morris, to house some of his employees. However, in view of the commanding position which it occupies and its remoteness from the copperworks and collieries, it must have been sited more with a view to creating a landscape feature than for the convenience of its occupants.

It is built of local sandstone, embellished with string courses of copper slag and some brickwork. It originally consisted of four square corner towers connected by lower ranges to form a quadrangle, but now only fragments of the two northern towers remain. Each tower contained a basement and three storeys. Traces of plas-

158 Forge Row, Clydach, prior to 'improvement'.

tering remain on some of the inner walls and these show up the outlines of some of the staircases. To the north of the building earthen banks divide the ground up into a number of small plots: it has been suggested that these mark out the potato patches of the original occupants.

An early source states that forty families lived in the castle, but an examination of the remains suggests that twenty would be more likely. It was still occupied in 1814, and may have been in 1850, but by 1880 it had been abandoned and was in ruins.

157   MORRISTON
★★   N I   [SS 6692 9767]   CA

The nucleus of this district of Swansea was a planned settlement created by John Morris with the object of providing housing for his workers. It was laid out on a grid plan in the 1780s by William Edwards, who is now chiefly remembered for his bridge at Pontypridd, but who in fact spent much of his life in Swansea [cf 153]. Morris did not actually have the houses built himself, but granted plots of a generous size on favourable terms to his employees who then erected their own houses. The focal point of the new township was St John's church, built in 1789. The present structure is the work of R K Penson, who also built Llandybie limekilns [24], and was opened in 1857. It ceased to be used for worship in 1970.

Housing construction started in the early 1790s and by 1796 there were said to be 141 dwellings in existence. Since then Morriston has been almost entirely rebuilt—in some areas more than once—and only two houses can be identified as belonging to the first period of construction, 35 Morfydd St and 91a Woodfield St, and both of these have been modified. The ruins of the market of 1827 remain on Market Street.

158   FORGE ROW, CLYDACH
★   I   [SN 6872 0188]   □

Only the shell remains in 1980s guise of what were once very attractive late 18th or early 19th century stone cottages with pantile roofs. In common with many workers' cottages, these had windowless rear elevations. They overlook the ravine that carried the tailrace from the attendant iron forge [39].

159 COLLEGE ROW, YNYSGEDWYN
★★　　N I　　　[SN 7858 0994]　　□

The building of this large sweeping curve of ironworkers' housing beside the River Tawe accompanied the phenomenal growth of Ynysgedwyn ironworks after 1837 [42] when the labour force totalled around 1000. Although the fenestration has been modernised with unfortunate effect, most of the houses retain their proportions and one or two survive largely in their original condition.

160 GOUGH'S ROW, YSTRADGYNLAIS
★　　N I　　　[SN 7870 1052]　　□

A row of houses just to the north of Ystradgynlais on the A4067 road. Their rear extensions are virtually built into the embankment which carries the road—originally the formation of the Swansea Canal. They were built to house workers in the collieries of the landowner (the Rev Fleming Gough) in the 1820s at a time when the village of Ystradgynlais hardly existed. George Crane, ironmaster of Ynysgedwyn [42] had to build his workers' houses here as an extension to the main row. These two-storeyed houses are now being improved out of recognition.

161 SHEPHERDS' COTTAGES, CAE'R-BONT
★★　　　　　　[SN 8019 1165]　　□

A distinctive group of stone-built cottages with elegant barge boards, chimneys and lattice windows. Built in the mid 19th century by the Aberpergwm Estate for workers on their Gwaunclawdd Farm. A detached house of similar design [SN 8018 1167] stands alongside the public footpath which leads to the circular byre at Gwaunclawdd.

162 PRICE'S ROW, COELBREN
★　　N I　　　[SN 8447 1134]　　□

An isolated terrace built by the railway contractor John Dickson in the 1860s when he was working on the Neath & Brecon Railway [135]. Probably intended initially to house some of his navvies. The road on which the terrace stands is the course of the Brecon Forest Tramroad [110-6] which Dickson purchased in 1865. The present name is not original, the terrace having been known first as 'Dickson's Row'.

163 CHURCH STREET, SEVEN SISTERS
★　　　　　　[SN 8177 0800]　　□

An interesting terrace of single-storey colliers' cottages on the main road into the village from the south. Commercial deep-level mining in the area started in 1872 and the village developed rapidly in the following decades. The name is said to derive from the seven sisters of Evan Evans Bevan, whose father sunk the first pit.

164 CYNONVILLE GARDEN VILLAGE
★　　　　　　[SS 8250 9520]　　□

Promoted by the Cynon Colliery Co in 1910 along the lines of Bournville or Port Sunlight as an incentive to attract labour to a hitherto unworked part of the coalfield. The original plan was to build 139 houses, nearly all semi-detached, with shops and recreational facilities, on a sloping site on the side of the Afan valley. The outbreak of war prevented the completion of the scheme and only the two lowest rows of houses were built.

165 BRICK STREET, GLYNCORRWG
★　　　　　　[SS 8755 9915]　　□

A street of three-storeyed houses built in the 1860s for officials of the Glyncorrwg Colliery Co and the South Wales Mineral Railway [140–42]. It was the opening of the railway in 1863 that made it possible to exploit this remote corner of the coalfield, and this in turn led to intensive house building in a village that had hitherto consisted of no more than a church and a few cottages. Interestingly, this and other early streets in Glyncorrwg were constructed by a Wolverhampton building firm, which reflects the fact that it was capital from the Black Country that built the railway and opened up the colliery.

For further sites at which domestic housing may be seen, see entries 40, 43, 45, 112, 115, 167, 168 and 170.

MILLS AND WATER POWER

Much of the industrial revolution in Wales was powered by water rather than by steam, and many of the earlier industrial works re-used corn mills for this reason. Later, elaborate water leats were constructed that ran along valleys for considerable distances and even crossed them on stone arched aqueducts, as at Llangyvelach copperworks in 1717 [26], Taibach copperworks in 1774 or the appropriately named Bont Fawr (Great Bridge) at Pontrhydyfen in 1824 [168]. The water wheel at Oakwood ironworks, with a diameter of 45' (13·7m) and a width of 10' (3m), which provided the furnace blast and developed 90 hp, was said to have been the most powerful in Wales at the time.

There are remains of two of these substantial leats in the Clyne Valley Country Park at Swansea. That on the eastern side of the valley was built c1800 to power a colliery [SS 6152 9178] and the New Mill [SS 6107 9227] which was also in the same hands as the colliery. In response the Duke of Beaufort, the seignorial lord, built the Clyne Wood Canal on the opposite side of the valley in order to provide an improved supply of water to his ancient manorial mill at Blackpill.

The water economy of the area grew ever more complex throughout the 19th century, so that the Swansea Canal and its feeders, for instance, were supplying water to drive over 40 installations at the same time as its most intensive use for navigation.

166   **FELINDRE WATERMILL**
☆   **N PS**   [SN 6375 0272]   **LS**

A fairly complete cornmill surviving in one of the more rural areas of the region.

167   **GLYNNEATH WOOLLEN MILL**
★   **N I**   [SN 8473 0615]   **CA**

A long range of mid 19th century buildings just to the south-west of Glynneath, comprising the manager's house, ten mill-workers' cottages and, at the north-east end, the mill itself. The waterwheel was operated by by-pass water from a lock on the Neath Canal. This was not the only use of water from the Neath Canal as a source of power, but the intensity of such use was much less than on the Swansea Canal.

168   **BONT FAWR AQUEDUCT PONTRHYDYFEN**
★★   **N PS**   [SS 7954 9415]   **SC**

A huge stone aqueduct, whose name means 'great bridge', built 1824-27 by John Reynolds to convey water at a suitable height to drive the waterwheel that provided the blast at his Oakwood ironworks. Its length as 459' (140m) and its height 75' (23m). After 1841 a railway was laid over the bridge and according to local tradition it was also used for small boats. The four 70' (21m) elliptical arches now carry a minor road.

For further sites at which the use of water power was a feature, see entries 1, 9, 34, 43, 45, 50, 59, 70, 88, 118, 119, 169 and 173.

168 Bont Fawr Aqueduct, Pontrhydyfen.

AGRICULTURE

Most of the population of the inland part of south-west Wales in the 18th and early 19th centuries was dependent on agriculture for their livelihood. The enclosure of the Great Forest of Brecon (Fforest Fawr) in the 1810s and 1820s prompted the further construction of limekilns and agricultural tramroads as an aid to improving the land. Agrarian ideas from Scotland and the Borders prompted the building of model farms such as the huge Forest Lodge with its water-powered threshing and corn-grinding machinery. Another Scottish innovation was the introduction of circular cattle byres with double outer walls enclosing an annular feeding passage: examples can be seen at Gwaunclawdd [SN 8101 1225] and Blaenpelena [SS 8251 9489]. One model farm at Belfont had a covered horse gin but these are unusual in this region.

The Lincolnshire banker Joseph Claypon took over the Crown Allotment of the Great Forest in the 1830s. His family's fortune had been made in draining the Fens and he spent another in driving drainage dykes up mountains in the Brecon Beacons. This scheme included two huge rabbit farms at Cefn Cûl and Pant Mawr [SN 8730 2030; SN 8950 1520] to supply pelts to workers in the new industrial towns. The long earthen rabbit warrens (or pillow mounds) and drainage dykes can be seen clearly from the A4067 road when passing Cray reservoir.

169 **FOREST LODGE**
☆☆   **N I**   [SN 9595 2425]   □

The Great Forest of Brecon was a huge tract of high moorland pasture in the western part of what is now the Brecon Beacons National Park. It was sold by the Crown to raise funds for the Napoleonic Wars, and three great farms were constructed by the purchasers here and at Belfont [170] and Cnewr [112] on the 1200' (366m) contour.

William Rowland Alder, an improving landlord from near Berwick-on-Tweed, bought about 1775 acres of the Great Forest. He planned wheat farms at this considerable altitude, protected by huge windbreaks, in a scheme whose costings were based on the high prices brought about by the war. Unfortunately prices collapsed before he could benefit from them. He repeatedly mortgaged his properties and eventually had to sell up.

Forest Lodge is a most impressive site. Huge barns had a waterwheel attached to drive threshing and corn-grinding machinery: the millstones remain on site. Other ranges of buildings include a manager's house and a courtyard of workers' houses, enlarged in the later 19th century. A mill dam also remains.

169 Mill at Forest Lodge Model Farm.

Gwaunclawdd Byre.

170 BELFONT
☆ **N I** [SN 8789 2565] ☐

A model farm built 1817-19 by William Alder [169]. A huge granary block remains at the rear of the original farm courtyard. The semi-circular horse-gin house that drove the threshing machinery in the granary has been demolished.

For further sites with agricultural remains, see entries 16, 17, 112, 161 and 166.

MISCELLANEOUS INDUSTRIES

171 BLACKPILL PYROLIGNEOUS ACID WORKS
☆ **N** [SS 6167 9069] ☐

Built in 1856 to use wood from the Clyne Valley. Wood was heated in large cast-iron cylinders and the resultant gases led through traps to a condensing worm and thence to a receiver. The resultant pyroligneous acid (an impure acetic acid) was largely consumed by manufacturers of acetate mordants for the dyeing trade. Tar and charcoal were by-products of the process. The business appears to have failed by 1863. A re-roofed half of the works serves as the garage of 'Mill Leat' in Blackpill Road and the adjoining 'Woodland Cottage' was an early domestic conversion of the north-eastern end of the building. The ruinous section in between gives more of an idea of the original character of the works.

172 PORT EYNON SALTHOUSE
★★ [SS 4694 8463] ☐

The remains of a 16th century (or even earlier) saltworks and associated fortified house on the foreshore at Port Eynon. A large stone chamber below the present level of the beach was the collecting point for salt water which was then pumped to a higher level where it was evaporated by artificial heat to leave a residue of salt. Another saltworks stood on the foreshore at Port Tennant, Swansea.

173 GLYNNEATH GUNPOWDER WORKS
★★ [SN 9184 0843] ☐

Established 1857 on the site of the former Dinas fire brick works. Ownership subsequently passed to Nobel's Explosive Co and finally to ICI. The product was black powder and when this was removed from the Home Office's permitted list there was no alternative but to close the works. This took place in 1931. The site was evacuated and deliberately fired in 1932 as a safety measure. The remains of the works stretch for nearly two miles along the banks of the River Mellte, and include the gutted buildings, leats which provided water power to drive machinery, and the horse-worked tramway which served as the central spine of the whole undertaking.